THE SUPERIOR
BRAIN OF WOMEN
AND CALCULUS

THE SUPERIOR BRAIN OF WOMEN AND CALCULUS

Olivia F. Snyder

Library of Congress Control Number: 2007906615
ISBN: Hardcover   978-1-4257-7551-3
               Softcover   978-1-4257-7536-0

This book was printed in the United States of America.

To order additional copies of this book, contact:
Xlibris Corporation
1-888-795-4274
www.Xlibris.com
Orders@Xlibris.com
41675

DEDICATION

For Jorge, Carol, and Victor

CONTENTS

Acknowledgments

I am indebted to the hundreds of authors whose insights, as disclosed in their books, have clarified for me a subject many people dread the most: mathematics, especially calculus.

Thanks to the guidance of renowned mathematicians, you will be able to unravel the mysteries of calculus. My perseverance and dedication to demystify what stubborn, egocentric, and arrogant math teachers have always done to calculus—to purposely confuse students with the sole aim to make the subject unclear and portray themselves as Einstein's direct descendants—will help you master calculus, which is really just very advanced algebra, and, at the same time, show you how easy and fascinating it really is.

I am also grateful to my family, my literary agent, and my publishing company's staff for all their support and guidance.

ø
Phi

| = | ∫ | + |
| Equation | Integral | Addition |

√
Radical

M²
Math Square

π
Pi

÷
Division

Σ
Sigma

∞
Infinity

Introduction

The title of this book, *The Superior Brain of Women and Calculus*, is based in the correlation between the function of women's brain and calculus.

Calculus, which is divided into two intrinsically linked areas: the *differential* and the *integral*, operates in its optimum capacity when the two areas are combined. In the same way, the human brain operates in its maximum capacity when the two hemispheres are used and connected simultaneously, as is the case of women who have *superior brains by being able to connect and use* the two hemispheres of the brain simultaneously, thanks to the more developed corpus callosum, in comparison to men.

In calculus, the differential is involved with analysis, as is the case with the left hemisphere of the brain; and the integral is involved with synthesis, as is the case with the right hemisphere of the brain, which has the power of integration, of visualizing the whole picture.

Corpus callosum is a structure of the mammalian brain in the longitudinal fissure that connects the right and left cerebral hemispheres. It is the largest white matter in the brain, consisting of 250 million contralateral exonal projections. Much of the interhemisphere composition of the brain is conducted through the corpus callosum.

Knowledge is power, and for that reason, women were deprived for thousands of years of that human intrinsic need. Without knowledge people become dependent on other people, and the learned can easily manipulate the ignorant.

For thousands of years, women were confined in their own homes and denied access to higher education. For all this time, universities were established exclusively for males, and in some parts of the world, women are still without access to develop their intellect. They were forced to live a life of doing menial, unimportant tasks, toiling for others, isolated, and bearing children year after year. In the past, millions of intelligent, knowledgeable, independent women were burned at the stake by the Catholic Church; and more recently, a group of *girls were burned alive for attending a school by another bestial religious group in an Islamic country.*

Around 415 CE (common era), Hypatia—perhaps one of the most famous mathematician and philosopher women that made a remarkable contribution in the development of mathematics, and who most probably occupied the chair of Neoplatonic philosophy at Alexandria—was brutally murdered by a Christian group supported by Alexandria's patriarch, Cyril, who was later on upgraded to Saint Cyril by the Catholic Church. She was not only a remarkable mathematician and scientist, but also a philosopher. Hypatia had advanced so far in her education that she surpassed by far the philosophers of her time and was able to address even the rulers. Because of her extraordinary knowledge and nonreligious beliefs, the Christian monks seized her on the street, beat her, and dragged her body to a church, where they mutilated her flesh with sharp tiles and later burned her body. After the brutal assassination, they proceeded to burn the Library of Alexandria—one of the most important libraries of all times—which was guarded by Hypatia and contained invaluable, original ancient texts, as well as books of that time. It was an irreparable and irreplaceable loss for humanity!

Patriarchal societies have been established throughout the world for more than two thousand years and still are prevalent. Nevertheless, groups of courageous, intelligent women around the world have been able to fight back and to open the door for other women to get access to education and public life. In the United States, thanks to the tireless, active, and enthusiastic action of NOW (National Organization for Women), founded in 1966, women have been able to enjoy their educational, political, and reproductive rights. This magnificent organization has been able to preserve and expand women's rights for decades.

Recently, in Iceland, girls broke all records completely and outshone Icelandic boys by a significant margin on all parts of a math test. Also, Japanese girls have scored higher in math tests overall than the boys of many other nations, including the United States.

To understand ourselves, we have to understand the universe. We all have been enchanted by its complexity and mysteries. Contemplating a myriad of different forms and rhythms, we see patterns, relationships, and signs. Through this understanding, we will eventually find a clue or a formula that is a key to discover the unifying principle—the wholeness.

Using the language of comparison and mathematical relationship, we disclose life's mysteries by placing the larger next to the smaller and holding them both up to the whole. What we then discover is a relationship of balance, harmony, and symmetry that is quite revealing.

The whole is to the larger in exactly the same proportion as the larger is to the smaller. It is easily described as a pattern of numbers that increases by adding the two previous numbers. For those of us who find numbers and abstractions difficult to understand, it means that there is a relationship that can be proven by numbers, that gives life to a series of shapes and dynamics that appear throughout nature. Furthermore, its principles of harmony have been acknowledged as fundamental truths in the world of spirit, and its relationship to our daily lives is seen to the proportions of our very own bodies.

The relation of macrocosm and microcosm describes the larger and smaller in their more intrinsic relationship. They are not separate entities; they are related.

Universal laws are acknowledged deep down in the consciousness as part of oneself and are eternally true. Throughout this book, we will learn different statements of its patterns—mathematical, creative, and spiritual.

This knowledge is not only important for our overall view of the universe, but it is also a vital tool to further our careers in the physical world.

Chapter 1

KNOWING THE OBJECT OF MATHEMATICS

Mathematics

Mathematics comes from the Greek word *mathematike*, which means "knowledge," "science." It discovers and maps out the real world—all existing things viewed as a whole—through its science dealing with quantities, forms, etc., and their relationship by the use of numbers and symbols. Its main objective is to prove theorems, which are axioms (assumptions) based on proofs (reasoning); and using its language, we are able to make the invisible visible.

Mathematical symbols, called numbers, serve two purposes. One is their practical use as tools to count and measure, and the other is the means in which we can try to understand the mysterious and the unexplainable universe. Numbers can be described as a universal language.

Above all, mathematics is based on **reasoning**. Therefore, it is involved in concepts related with reason, such as:

Ratio and **rate** are derived from the Latin words *ratio* and *onis*, which signify reason, and *reri* and *ratus*, meaning "to think and judge." In mathematics they both mean *the measurable relation of one thing to another (e.g., a dollar **per** item) or the relation of quantities as shown by their **quotient** or **proportion***. Ratio can be expressed as a *division* or *fraction*.

Rational is derived from the Latin word *rationalis*, which signifies **reason**. In mathematics a **rational number** is any number that can be expressed as a **ratio** of two numbers with no decimal, namely, a whole number, or with a decimal that has an ending. For example,

$$9 \div 3 = \textbf{3}, \text{ or } 6 \div 5 = \textbf{1.2.}$$

Note: $9 \div 3 = 3$ is **exactly the same** as $\sqrt{9} = \textbf{3}$, therefore, a **rational number** is also any number that can be expressed as a ratio of a **squared number** with no decimal, namely a whole number, as it is in this particular case.

Irrational is derived from the Latin word *irrationalis*, which signifies lacking, the power to **reason**. In mathematics an **irrational number** is any number that cannot be expressed as a **ratio** of two numbers with a decimal without end. For example,

$$\sqrt{3} = 1.73... \text{ or } \sqrt{2} = 1.41...$$

Basically, mathematics connects patterns of growth with natural laws. Through the endless applications of properties that emerge from its geometry, we are able to see proportions and symmetry all around us, from the creation of our bodies to the spiral patterns of the universe.

In mathematics, we are able to find ways to describe order through numbers in what at first appeared to be chaos, to collect empirical data about ourselves and the universe as a whole.

We are all amazed by the complexity of the universe. We all agree that life is endlessly mysterious. Mathematics gives us a way to gaze at a myriad of different forms and rhythms and contemplate infinite patterns and relationships. In essence, we can describe mathematics as follows:

Mathematics

↙   ↘

Mathematics of unchanging rates   *Mathematics of infinitesimal changes*

algebra, geometry, trigonometry   calculus

differential calculus

Process of finding the *slope of a curve*

(rate, derivative—measure of comparable changes)

|

integral calculus

Process of *finding* area of the whole

and a finite sum

(through *addition* of *infinitesimal*

number of numbers)

In ancient times, people started to count such things as their herds and flocks. Most probably, they used their fingers or pebbles to keep track of small numbers. Numbers are also called digits, which mean fingers.

Brief Story of the Development of Mathematics and Physics—Mathematical Description of Nature

By 3000 BCE, the people of ancient Babylonia, China, and Egypt developed a practical system of mathematics. They used written symbols to stand for numbers and knew simple arithmetic operations that they put in practice in business and government. They also developed a practical geometry helpful in agriculture and engineering. Egyptians knew how to make the intricate measurements necessary to build the pyramids and to survey their fields. Babylonians and Egyptians even started to explore some of the fundamental ideas of algebra.

Furthermore, scholars have translated clay tablets that show that the Babylonians were highly skilled in arithmetic and astronomy more than 4,000 years ago. They developed the system we use today **to measure angles and degrees, minutes, and seconds**. Because there are sixty seconds in a minute and sixty minutes in an hour, this system is based on tens up to sixty and on sixties from here on. The clay tablet also showed that about 2,400 years ago, the Babylonians had a **symbol for zero** and another symbol that worked in the same way as a **decimal point**. Although we inherited the idea of time and angles, the Babylonian idea of place value was lost until the Hindus rediscovered it.

The Hindus elaborated our present-day numeral system, and the Arabs brought it to Europe sometime before 1200. However, it was not until 1600 that the decimal point and decimal fractions were introduced.

Between 600 and 300 BCE, the Greeks took the next great step in mathematics. They inherited a large part of their mathematical knowledge from the Babylonians and Egyptians. They separated geometry from practical applications and made it into an **abstract exploration of space**. They based this study of points, lines, and figures—such as triangles and circles—on logical reasoning rather than in facts found in nature.

Before Common Era (BCE)

Thales of Miletus (c.640-546 BCE), Pythagoras (c.580-c.500 BCE), Euclid (c.300 BCE), and many other Greek mathematicians organized geometry as a single logical system. As early as 450 BCE, Greek mathematicians recognized **irrational numbers** such as the square root 2. Aristarchus de Samos (c.310-230 BCE) Greek mathematician and philosopher was the first person who proposed the heliocentric model of the solar system in the center of the universe. He realized that the sun was larger than the earth and that the earth orbited a motionless sun. Archimedes (c.287-212 BCE) devised processes that foreshadowed those of **integral calculus.**

Common Era (CE)

Ptolemy (c.150 CE) helped developed trigonometry. Diophantus (c.275 CE) helped found **algebra. Hypatia of Alexandria** (c.370-415 CE), an outstanding mathematician, scientist, and philosopher, who was the first woman to make a remarkable contribution to the development of mathematics. As stated before, she had progressed so far in her education that she surpassed by far the philosophers of her time and was able to address even the rulers. This was too much of a challenge for a sect of Christians, who savagely murdered her.

The **Middle Ages was** the period in Western European history between the fall of the Roman Empire and the dawn of the Renaissance period—roughly the fifth through the fifteenth centuries CE. The centuries preceding the eleventh century are called the **Dark Ages.** People were reduced to subsistence level. During this period, Europe saw no new developments in mathematics for hundreds of years. But the Arabs preserved the mathematical tradition of the Greeks and Romans. Mathematicians in India further developed **zero** and the **decimal number system** taken from the Babylonians, who invented them about 2,400 years according to ancient clay tablets found by archaeologists as stated before. Around 820 CE, the Arab mathematician Al-Khowarizmi organized and expanded algebra. The word *algebra* comes from an Arabic word that he used in the title of one of his books. After 1100 CE, Europeans started to borrow the mathematics of the Arab world, and scholars started to study algebra and geometry. In 1200 CE, Leonardo Fibonacci, a renowned mathematician, contributed to arithmetic, algebra, and geometry.

The Renaissance, from the twelfth to the sixteenth century, produced great advances in mathematics. The invention of the printing press brought the publication of arithmetic textbooks. Many of the computation methods used today date from this period. Michael Stifel (1487-1567), Niccolo Tartaglia (c.1500-1557), Girolamo Cardano (1501-1576), and Francois Viete (1540-1603) pioneered algebra. Viete introduced letters to stand for numbers.

But the most extraordinary events started to happen at the time when Nicolas Copernicus (1473-1543), a Polish astronomer, disproved the absurd idea of the Catholic Church that the earth is the center of the universe. He also contributed to mathematics through his research in astronomy. And later, **Bruno Giordano** (1548-1600), a remarkable Italian philosopher, taught the infinity of the universe and the truth of the Copernican hypothesis that the earth was not at the center of the universe. **For his courage, clarity of thought, and intelligence, and because he openly confronted the ignorance, stupidity, and stubbornness of the Catholic Church, he was burned at the stake.**

The 1600s also brought many brilliant contributions to mathematics and the perception of the universe: John Napier (1530-1617), a Scottish mathematician, invented logarithms. **Galileo Galilei** (1564-1642), an Italian mathematical physicist, discovered the laws of falling objects and the parabolic motion of projectiles. He was also a talented publicist, who helped to popularize the pursuit of science, which had always been seen as an enemy of the Catholic Church.

Many churchmen opposed him, but Galileo firmly upheld the theory of Copernicus that the earth moved around the sun. Church officials warned him to abandon the Copernican system, and at the same time, the church placed the work of Copernicus on the index of prohibited books, where it remained for 200 years.

In 1632, Galileo published his masterpiece, *A Dialogue on the Two Principle Systems of the World.* The Holy Office, or Inquisition, immediately called him to appear before it. After a long trial, church officials forced him to say that he gave up his belief in the Copernican theory and sentenced him to an indefinite prison term. During his last years, he spent his time writing on the laws of force and motion. He summed up his life's work on motion, acceleration, and gravity in his book *Dialogues in the Two New Sciences*, published in 1638; and furnished a basis for the three laws of motion laid down by Sir Isaac Newton in 1687. Before Galileo published his last book, he became blind, but the Inquisition always kept an eye on him.

Therefore, Galileo and Newton, along with Johannes Kepler (1574-1660), expanded mathematical knowledge through their studies of the stars and planets. Rene Descartes (1596-1650) invented analytical geometry. Pierre de Fermat (1601-1665) and Blaise Pascal (1623-1662) invented the mathematical theory of probability. Then toward the end of this period, Sir Isaac Newton and Gottfried Wilhelm Leibnitz (c.1675) furthered the idea of **calculus**. Other great contributors include Leonhard Euler, Joseph Louis Lagrange; and later on there were Augustin L. Cauchy, Karl F. Gauss, Pierre-Simon, Adrien-Marie Legendre, Luitzen Brouwer, George Cantor, David Hilbert, Bertrand Russell, and Alfred North Whitehead.

Emile Du Chatelet (1706-1749) was a remarkable scientist and mathematician Frenchwoman who corrected Voltaire and improved on Newton. She found that:

Kinetic energy is equal to mv^2; that is, $Ek = mv^2$
(kinetic energy (Ek) equals mass (m) multiplied by the
square of its velocity (v)).

In this way, this extraordinary scientist and mathematician contributed greatly to the concept of Albert Einstein's special theory of relativity:

Energy is equal to mc^2; that is, $E = mc^2$
(energy (E) equals mass (m) multiplied by the square of
the speed of light (c)).

Take note that Du Chatelet formula has been shown later to be $Ek =(1/2)\, mv^2$ where Ek is the kinetic energy of an object, m its mass and v its velocity. Also Du Chatelet complete name was: Gabrielle Emilie Le Tonnelier De Breteuil, Marquise Du Chatelet.

Although it is obvious that v stands for velocity, which means quickness of motion from the Latin word *velocitas—velox*, swift. There is another velocity factor in Einstein's formula represented by c which equals celerity meaning quickness, rapidity of motion from the Latin world *celeritas—celer*, quick. As we can clearly see both words, velocity and celerity have the exact meaning. Kinetic energy means energy derived from motion from the Greek word *kinetikos—kineein*, to move, while energy means capacity of acting or being active from the Greek word *energeia—en*,in, *ergon*, work. Einstein's equation deals with rest mass energy.

In 1905 three articles were published by **Albert Einstein** and signed in the original "Einstein-Maric" announcing the general theory of relativity and other works. This important event was witnessed by the Russian physician A. Joffe (or Loffe). Christopher Jon Vjerknes, author of *Albert Einstein: The Incorrigible Plagiarist*, defends Mileva Maric in backing up proofs that she actually contributed significantly to the work published under the name of Albert Einstein only, which later won him the Nobel Prize. **Mileva Maric (1875-1948),** an intelligent mathematician Serbian woman, was Einstein's first wife (1903-1919), who actually contributed greatly in his work. They had a marital agreement to publish their work under both of their names: Einstein and Maric. This agreement has been confirmed later by several letters written by both spouses, which is part of Einstein's biography. **Einstein left his family, erased Maric's name, and was awarded the Nobel Prize in 1921 with his name alone.** For further information on this subject, visit www.pbs.org and access its documentary entitled *Einstein's Wife: The Life of Mileva Maric Einstein*, which can be bought online or by calling toll free 1800-play-PBS.

Albert Einstein also believed that the celebrated formula $E = mc^2$ might explain the theory by Marie Sklowdowska-Curie (1867-1934), a Polish physicist who discovered radioactivity (radioactive elements of polonium and radium). She found out that just an ounce of radium emitted 4,000 calories of heat per hour indefinitely. She won two Nobel Prizes in 1903 and 1911. Marie Sklowdowska-Curie shared her first Nobel Prize with her husband, companion, and assistant, Pierre, because of the great love she felt for him; but she was actually the *only* discoverer as she stated on several occasions.

In 1938 Lise Meitner, an outstanding Austrian physicist, discovered the fission of heavy nuclei, verifying Einstein's special theory of relativity.

She is considered one of the most significant scientists of the twentieth century. She realized that the uranium nucleus had split rather than merely having small particles chipped off it, and she was the first person in the world to explain what had happened. But she was not given credit for her discovery; instead, credit was given to

Otto Hahn, **who plagiarized her work,** and took **sole credit,** ignoring Meitner's discovery and her arduous collaboration with him. As a result, in 1944, the Royal Swedish Academy of Sciences awarded Otto Hahn the Nobel Prize in Chemistry for his supposed "discovery of the fission of heavy nuclei"; and Meitner was ignored.

Of course, in Meitner's case, there were a myriad of political and personal reasons that were involved in not given her the Nobel Prize. But in spite of all the reasons that we can find throughout the Internet or other sources why she was not awarded, the facts and true cannot be changed and history will give her full credit for her work.

Besides, throughout history, we have seen that in our patriarchal society it looks better when a man is awarded rather than a woman, so everybody was quite satisfied with the Royal Swedish Academy of Sciences' decision.

As we can see throughout our patriarchal socialization—and as argued by Gerda Lerner, author of the remarkable book *The Creation of Patriarchy*—the establishment consisted not of one event but of a series of events occurring over a period of nearly 2,500 years, from around 3100 to 600 B.C.E., at different paces and at different times in several societies; and women have had to struggle against men's imposition to gain access to their intrinsic right to education and for their intellectual development. Not only have men been able to ban access to educational institutions for hundreds of years, but they have also plagiarized women's work on an unknown number of occasions throughout history.

Symbols—Important Tools to Understanding Our World and the Universe as a Whole

Symbols are apprehended by the mind and reach the preconscious world of the spirit, the conscious, and the subconscious. They are all important to express our thoughts, emotions, and perceptions.

Around 800 BCE, when the Greeks came in contact with the Phoenicians, they borrowed their symbols to make their own alphabet. The Phoenicians used more consonants than the Greeks, so they used the extra signs for the vowel sounds. That was an improvement over the Egyptians' and Phoenicians' systems because it was possible to combine both consonants and vowels to create any sound they needed. Later on, the Romans adopted this type of alphabet and eventually developed it into much the way we use it today.

It was not until the fourteenth century that the use of plus and minus symbols (+ and -) first appeared in text. However, these symbols were in use long before they appeared in mathematical texts. The equal sign (=), composed of two parallel signs, was not used until 1557.

Understanding the Mathematical System

The mathematical system is the fundamental interdisciplinary tool encompassing all science, which is any systematic field of study or body of knowledge that aims—through observations, experiment, and deduction—to generate reliable explanations of the material and physical world phenomena. It can be divided into two branches: **pure** and **applied mathematics**.

Pure mathematics has as its basis the abstract study of quantity, order, and relation. It includes concrete numbers (*arithmetic*) and abstract numbers (*algebra*, *geometry*, and *calculus*). **Applied** mathematics deals with the application of this abstract science, especially in physics and engineering.

The first things we learn in school are an assortment of written symbols to stand for letters and numbers. A group of letters form words to express an idea. Numbers are a collection of counting words expressed in written symbols called numerals, and they comprise the tools to count and measure; at the same time, they are an important vehicle to transport us to visualize and understand the unexplainable and mysterious cosmos. Our present numerical system, elaborated by the Hindus and brought to Europe by the Arabs, is sometimes called the Arabic or Hindu-Arabic numeral system. It is also called the **decimal numeral system**.

In order to understand basic as well as advanced mathematics, we have to understand the meaning of all its symbols.

Symbols
The Number System

Our number system has different types of numbers (symbols): **concrete and abstract**. The most basic ones are **concrete numbers**: *natural numbers*—integers/whole numbers, prime numbers, composite numbers, as well as *real numbers*—rational and irrational numbers for basic **arithmetic**. **Algebra** and advanced algebra, or **calculus,** deal with sets of **abstract numbers** on which certain formal operations are defined.

Arithmetic *Concrete* Numbers

Real Numbers
Real Numbers have no imaginary parts, such as:

Natural Numbers

1, 2, 3, 4, 5, 6, . . .

Whole Numbers

0, 1, 2, 3, 4, 5, . . .

Integers

... -3, -2, -1, 0, 1, 2, 3 ...

Prime Numbers
factors itself and 1
no product of 2 factors

1, 2, 3, 5, 7, 11, 13, 17, 19, 23, 29, 31, 37, 41, 43, 47 ...

Composite Numbers
neither *factor* is 1
product of 2 factors

4, 6, 8, 9, 10, 12, 14, 15, 16, 18, 20, 21, 22, 24, 25 ...

Rational Numbers

Any real number that can be expressed as a ratio of two integers (positive and negative whole numbers), providing the second number is not zero.

Fractions		Decimals
1/4	=	0.25
... 8/16 = 4/8 = 2/4 = 1/2	=	0.50
3/4	=	0.75
... 16/16 = 8/8 = 4/4 = 2/2 = 1/1	=	1.00 Integer
1/5	=	0.20
2/5	=	0.40
3/5	=	0.60

4/5	=	0.80
5/5	=	1.00 Integer
... 12/24= 6/12 = 3/6 = 1/2	=	0.50
6/6	=	1.00 Integer

Fractions   Repeating Decimals

1/6	≈	0.1666 ...
... 8/24= 4/12 = 2/6 = 1/3	≈	0.333 ...
... 16/24= 8/12 4/6 = 2/3	≈	0.666 ...
5/6	≈	0.8333 ...

*   **Remember,** *a fraction* is **exactly** the same thing as *a division*:
Its <u>denominator</u> = <u>divisor</u>, its <u>numerator</u> = <u>dividend,</u> and the number
of times one quantity is contained in another = quotient (ratio).

Irrational Numbers

Any real number which cannot be expressed as a ratio of two
integers. Irrational numbers have decimals that go on forever and
never repeat, as in the case of some square roots. For example,

Square Roots

$$\sqrt{2} \approx 1.41421356237 \ldots$$
$$\sqrt{3} \approx 1.7320508075 \ldots$$

Imaginary Numbers

The *imaginary numbers* are square roots of negative numbers. All of them are *real numbers* multiplied by $i = \sqrt{-1}$. For example,

$$\sqrt{-49} = i\sqrt{49}$$

and then

$$i\sqrt{49} = 7i$$

Complex Numbers

The *complex numbers* are all possible sums of *real* and *imaginary numbers.* They are written as

a + bi, where a and b are *real* and $i = \sqrt{-1}$ is *imaginary.*

Algebra/Calculus *Abstract* Numbers/Symbols

Transcendental/Irrational

Golden Proportion

ø phi   ≈   1.61803398874 . . .
representing the proportion which is mathematically described as the relation, in perfect balance, of the whole to its parts.

| e | | $\approx$ | 2.7182818 ... * |

representing one of the limit of infinite series.

| π Pi | | $\approx$ | 3.14159265 ... * |

representing a symbol for the ratio of the circumference of the circle to the diameter. This symbol, π, is called *radian* and equals 180°. It also represents one of the limit of infinite series.

General Symbols

| a,b,c | | = | Representing a notation that should be |

consistent and unambiguous: the same symbol should denote the same thing whenever it occurs, and not be used for more than one thing.

Specific Symbols

| (x,y) | | = | *Pair of Coordinates symbol* |

representing x and y coordinates on the x-y coordinate system.

$\dfrac{dy}{dx}$ =   *Derivative symbol*

representing ratio of little bit of y (Δy) to a little bit of x (Δx).

Δx =   *Infinitesimal **Derivative** symbol* representing Δx as it approaches zero in the limit, $\dfrac{dy}{dx} = \dfrac{\Delta y}{\Delta x}$.

$f(x)$ =   ***Function** symbol* representing any function.

$\sqrt{}$ =   ***Radical** symbol* representing a symbol to find the *square* root of a certain number. *Square root* is found by multiplying a *number* by itself until it equals a *given radicand.*

$\int$ =   *Indefinite **Integral**—Antiderivative symbol* representing the adding up of small pieces to get the *total* of the *whole.*

$\int f(x)dx$ =   *Indefinite **Integral** of Function f(x)symbol* representing the *family of all antiderivatives* of the function.

$\int_{a}^{b}$ =   *Definite **Integral** Symbol* indicates to compute function area between **a** and **b**, which represents numbers called *limits of integration.*

$\int_{a}^{b} f(x)dx$ =   *Definite **Integral** of Function f(x)symbol* gives total area between **a** and **b** under some *curve f(x).*

$\sum$ Sigma   =   *Summation* Symbol
representing the summing up of long series of numbers.

Arithmetic, Algebra, and Calculus Relationship Symbols

=		equal sign, "is equal to"
≠		"is not equal to"
≈		"is approximately equal to"
>		"is greater than"
<		"is less than"
+		Addition
× • ( )		Multiplication
−		Subtraction
/ ÷ —		Division

Algebra Patterns

Circle   =   Unity, oneness and the source
The mother of all shapes

Circle area: $A = \pi r^2$
Circle circumference: $C = 2\pi r$

Triangle   =   For right triangle (i.e., a triangle with a 90° angle), **Pythagorean Theorem says that**

$$a^2 + b^2 = c^2$$

Other Symbols

N   =   Denotes any number

n!   =   "Factorial n" is defined to be the product of all positive integers equal to or less than n.

!   =   Factorials*

i =   Complex/pair number
$(x,y) = (1,0)$ or $(0,1)$

$-i$ = $(x,y) = (-1,0)$ or $(0,-1)$

∞ =   Infinity

"   Number e is the limit of infinite series: $1 + \dfrac{1}{1!} + \dfrac{1}{2!} + \dfrac{1}{3!} + \dfrac{1}{4!} \cdots$

Number π **pi** is the limit of infinite series: $\dfrac{4}{1} - \dfrac{4}{3} + \dfrac{4}{5} - \dfrac{4}{7} + \dfrac{4}{9} \cdots$

r stands for radius

Factorial symbols are all positive integers less than or equal to the given number. For example,

$$2! = 2 \bullet 1 = 2; \ 3! = 3 \bullet 2 \bullet 1 = 6; \ 4! = 4 \bullet 3 \bullet 2 \bullet 1 = 24;$$
$$5! = 5 \bullet 4 \bullet 3 \bullet 2 \bullet 1 = 120; \ 6! = 6 \bullet 5 \bullet 4 \bullet 3 \bullet 2 \bullet 1 = 720.$$

When there are factorials in the numerator and denominator of a fraction, it is possible to cancel the same number. For example,

$$\frac{6!}{5!} = \frac{6 \bullet 5 \bullet 4 \bullet 3 \bullet 2 \bullet 1}{5 \bullet 4 \bullet 3 \bullet 2 \bullet 1} = 6 \qquad \frac{5!}{6!} = \frac{5 \bullet 4 \bullet 3 \bullet 2 \bullet 1}{6 \bullet 5 \bullet 4 \bullet 3 \bullet 2 \bullet 1} = \frac{1}{6}$$

$$\frac{(n+1)!}{n!} = n+1 \qquad\qquad \frac{n!}{(n+1)!} = \frac{1}{n+1}$$

$$0! = 1$$

Note: Prime numbers are **chief** numbers—divisible by no other number except 1 and itself. These general symbols, x and y, are combined by

Addition: (x+y)
Subtraction: (x−y)
Multiplication (x•y) or (xy)
Division (x:y) or (x/y)

Remember

− inverse +	÷ inverse ×
3 − 2 = 1 + 2 = 3	4 ÷ 2 = 2 × 2 = 4

Even numbers are divisible by **2;** if not, numbers are called **odd.**

For example, this fraction, $\frac{2}{3}$, reads "2 over 3" or "2 ÷ 3" or " 3$\overline{)2}$ ".

Order of Numbers

Decimals, Whole Numbers with Decimals, Whole Numbers

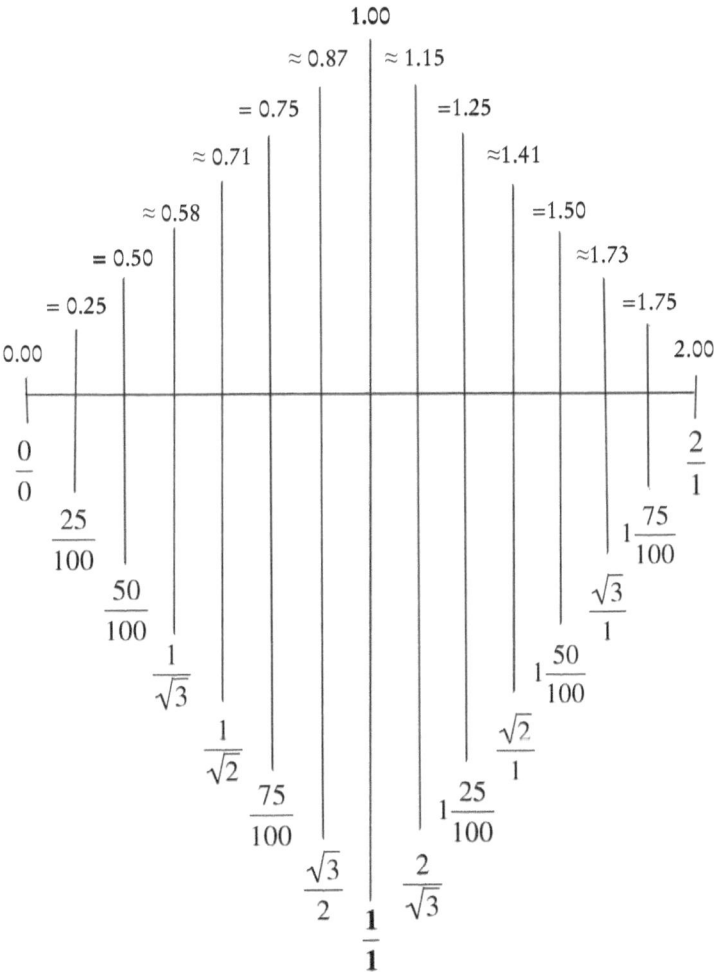

Fractions, Fractional Roots, Whole Numbers with Fractions

Note: $\dfrac{1}{\sqrt{3}} = \dfrac{\sqrt{3}}{3}$, $\dfrac{1}{\sqrt{2}} = \dfrac{\sqrt{2}}{2}$, $\dfrac{2}{\sqrt{3}} = \dfrac{2\sqrt{3}}{3}$, $\dfrac{\sqrt{2}}{1} = \sqrt{2}$, $\dfrac{\sqrt{3}}{1} = \sqrt{3}$

Order of Operations

Arithmetic operations (+, −, ×, ÷, and raising to powers) are *always* performed in specific order. However, expressions enclosed in parentheses are *evaluated first*:

1. Parentheses
2. Exponents
3. Multiplication and Division (left to right)
4. Addition and Subtraction (left to right)

Note: There is a mnemonic PEMDAS, which is expanded to the phrase "Please Excuse My Dear Aunt Sally."

Geometry

We need to master geometry in order to be able to measure the world. Geometry comes from the Latin and Greek word *geometria*—*ge* means earth, *metron* means a measure. Geometry is the branch of mathematics dealing with the properties, measurement, and relationship of points, lines, planes, and solids.

Coordinate Geometry

Given two points (x_1, y_1) and (x_2, y_2):

Slope

$$m = \frac{y_2 - y_1}{x_2 - x_1}$$

Distance

$$d = \sqrt{(x_2 - x_1)^2 + (y_2 - y_1)^2}$$

Midpoint

$$\left(\frac{x_1 + x_2}{2}, \frac{y_1 + y_2}{2} \right)$$

Geometry Table

Circle. A plane figure bounded by one line every point of which is equally distant from a certain point called the center.

$$Area = \pi r^2 \qquad Circumference = 2\pi r \text{ or } \pi d$$

$$Sector = \pi r^2 \times \left(\frac{\theta}{360°}\right) \qquad Arc\ length = 2\pi r \times \left(\frac{\theta}{360°}\right)$$

(where θ is the central angle)

Sphere. A solid body bounded by a surface of which all points are equidistant from a center.

$$Area = 4\pi r^2 \qquad\qquad Volume = \frac{4}{3}\pi r^3$$

Triangle. A three-sided figure on the same plane whose three angles always equal to 180°. Many calculus problems involve two right triangles. One is the 45°-45°-90° triangle with a shape of a 1 × 1 square cut in half along its diagonal and with a hypotenuse length of $\sqrt{2} \approx 1.41$ given by the Pythagorean theorem. Another is the 30°-60°-90° triangle with a shape of an equilateral triangle cut in half with a 2-unit long hypotenuse.

All Triangles: **Equilateral Triangle:**
(all sides equal)

$$Area = \frac{1}{2}base \times height \qquad\qquad Area = \frac{side^2\sqrt{3}}{4}$$

Parallelogram. A four-sided figure on the same plane whose opposite sides are parallel and equal.

$$Area = base \times height$$

Trapezoid. A plane figure bounded by four lines with two parallel sides.

$$Area = \frac{base_1 + base_2}{2} \times height$$

Square. It is an equilateral rectangle. A four-sided figure with four right angles lying on the same plane.

$$Area = s^2 \text{ (s = side length)}$$

Cone. A solid figure with a circular or elliptical base tapering to a point.

$$Volume = \frac{1}{3}A \times height \text{ (A = area of the base)}$$

Cylinder. A figure generated by a straight line remaining parallel to a fixed axis and moving round a closed curve and **prism.** A solid whose ends are similar, equal, and parallel polygons, and whose sides are parallelograms.

$$Volume = A \times height \text{ (A = base area)}$$

$$Area = P \times height \text{ (P = base perimeter/circumference)}$$

Trigonometry Formulas

Pythagoran Identity: $\sin^2\theta + \cos^2\theta = 1$

Half Angle:	Double Angle:	Reduction:
$\sin^2\theta = \dfrac{1}{2}(1 - \cos 2\theta)$	$\sin 2\theta = 2\sin\theta\cos\theta$	$\sin(-\theta) = -\sin\theta$
$\cos^2\theta = \dfrac{1}{2}(1 + \cos 2\theta)$	$\cos 2\theta = 2\sin\theta\cos\theta$	$\cos(-\theta) = \cos\theta$

$$\tan(-\theta) = -\tan\theta$$

Note: θ is the central angle.

Once we fully understand the meaning of *mathematical symbols* and *geometry*, we have to master the most important mathematical *theorems, formulas, and rules*, whose definitions are given throughout this book:

Six theorems

Fundamental theorem of algebra

Fundamental theorem of arithmetic

Fundamental theorem (first and second version)

Mean Value Theorem for Continuous Function

Mean Value Theorem for Integral /Average Value

Pythagorean Theorem

Sixteen formulas

Anti derivatives Formula

Arc Length Formula

Circumference Formula

Difference Quotient Formula

Integration of Parts Formula

Inverse Functions Formula

Limits Formula

Mathematics of Limits Formula

Quadratic Formula

Slope Formula

Speed Formula

Sum of Left Rectangles Formula

Sum of Midpoint Rectangles Formula

Sum of Right Rectangles Formula

Sum of Parabola Trapezoids Formula

Sum of Trapezoids Formula

Ten Rules

Chain Rule

Constant Rule

Constant Multiple Rule

Definite Integral Rules

Difference Rule/Sum Rule

Geometric Series Rule

L'Hopital's Rule

Power Rule

Product Rule

Quotient Rule

The Ten Most Critical Concepts of Calculus

1. Factor Patterns

$$a^2 - b^2 = (a-b)(a+b)$$
$$(a+b)^2 = a^2 + 2ab + b^2$$

2. Fractions

$$\frac{0}{5} = 0 \qquad \frac{5}{0} = undefined$$

3.   Powers

$$x^0 = 1 \qquad x^{1/2} = \sqrt{x}$$
$$x^{1/3} = \sqrt[3]{x} \qquad x^{2/3} = \sqrt[3]{x^2}$$
$$x^{1/5} = \sqrt[5]{x} \qquad x^{3/8} = \sqrt[8]{x^3}$$

Note: $\sqrt{x}$ is called square root, $\sqrt[3]{x}$ is called cubic root, $\sqrt[5]{x}$ is called fifth root, etc.

$$x^{-3} = \frac{1}{x^3} \qquad\qquad \left(\frac{1}{2}x\right)^2 = \frac{1}{4}x^2$$

$$\left(\frac{1}{2}\right)^3 = \frac{1}{8} \qquad\qquad \left(\frac{1}{2}\right)^4 = \frac{1}{16}$$

Differentiation:	Integration:
Power Rule	*Reverse* Power Rule

- Decrease *power* by 1
- Include *old power* in front

- Increase power by 1
- Divide new power

$$\frac{d}{dx}x^1 = 1 \qquad\qquad \int x^1 = \frac{1}{2}x^2 + C$$

$$\frac{d}{dx}x^2 = 2x \qquad\qquad \int x^2 = \frac{1}{3}x^3 + C$$

$$\frac{d}{dx}x^3 = 3x^2 \qquad\qquad \int x^3 = \frac{1}{4}x^4 + C$$

$$\frac{d}{dx}x^4 = 4x^3 \qquad\qquad \int x^4 = \frac{1}{5}x^5 + C$$

$$\frac{d}{dx}5x^4 = 4(5)x^3 = 20x^3 \qquad\qquad \int 20x^3 = \frac{20}{4}x^4 + C = 5x^4 + C$$

4. Roots

$$\sqrt{x} = x^{1/2}$$
$$\sqrt{0} = 0 \qquad \sqrt{1} = 1$$
$$\sqrt[4]{x^3} = x^{3/4} \qquad \sqrt{a^2 b^2} = ab$$

5. Differentiation $\dfrac{d}{dx}\pi = 0 \qquad \dfrac{d}{dx}3x = 3 \qquad \dfrac{d}{dx}kx = k^{*}$

* π is a number ≈ 3.14159265; k is a constant (any number).

6. Integral differentiation

$$\int x^2 = \frac{1}{3}x^3 + C \qquad\qquad \int \frac{1}{3}x^2 = x^3 + C$$

7. Logarithms

$\log_2 8 = 3$ is equal to $2^3 = 8$

8. Trig differentiation

$$\frac{d}{dx}\sin x = \cos x \qquad\qquad \frac{d}{dx}\cos x = \sin x$$

$$\frac{d}{dx}\tan x = \sec^2 x \qquad\qquad \frac{d}{dx}\sec x = \sec x \tan x$$

$$\frac{d}{dx}\cot x = -\csc^2 x \qquad\qquad \frac{d}{dx}\csc x = -\csc x \cot x$$

9. Trig functions   **Soh**   **Cah**   **Toa**

$$\sin\theta = \frac{O}{H} \qquad \cos\theta = \frac{A}{H} \qquad \tan\theta = \frac{O}{A}$$

Inverse:

$$\csc\theta = \frac{H}{O} \qquad \sec\theta = \frac{H}{A} \qquad \cot\theta = \frac{A}{O}$$

Note: For clarification:

Letter:	Stands for:
O	opposite side
A	adjacent side
H	hypotenuse

10. Trig values

$$\sin 30° = \frac{1}{2} \qquad\qquad \cos 60° = \frac{1}{2}$$

$$\sin 60° = \frac{\sqrt{3}}{2} \qquad\qquad \cos 30° = \frac{\sqrt{3}}{2}$$

$$\sin 45° = \frac{\sqrt{2}}{2} \qquad\qquad \cos 45° = \frac{\sqrt{2}}{2}$$

$$\tan 30° = \frac{\sqrt{3}}{3} \qquad \tan 45° = 1 \qquad \tan 60° = \sqrt{3}$$

Note: There are two ways to measure angles—*degrees* or *radians*. The *radian* measure of an angle is the length of the arc along the circumference of the *unit circle* cut off by the angle. To convert from *degrees* to *radians*, multiply the angle's measure by $\frac{\pi}{180°}$; and from *radians* to *degrees*, multiply the angle's measure by $\frac{180°}{\pi}$.

* The derivative of any number is zero, *unless* the constant or number is accompanied by a variable. In such a case, the derivative is the constant or number. Pi, π, is just a number (≈ 3.14159), not a variable.

The **slope definition** is also crucial in the concept of calculus.

Slope = ratio of the rise to the run

$$Slope = \frac{rise}{run} = \frac{y - coordinates}{x - coordinates}$$

Note: Rise is the vertical line called y-coordinate, and run is the horizontal line called x-coordinate.

Circle and Triangle

Circle and **triangle** are two key elements in the world of mathematics.

The **ci rcle**, in antiquity, was referred to as the *Oneness*—the first, the seed, the essence, the builder, and the foundation. They also called it *Unity*. It is the matrix, the womb in which anything is formed and from which all other forms emerge. *Matrix* comes from the Latin words *matrix*, "the womb," and *mater*, "mother." In mathematics a matrix is a rectangular array of quantities or symbols.

The **triangle** encloses the smallest area within the greatest perimeter in comparison with the **circle**, which encloses the greatest area within the smallest perimeter.

Chapter 2

REVEALING THE MYSTERY OF NUMBERS THROUGH ARITHMETIC

Arithmetic

*A*rithmetic (from Greek *arithmos*, meaning "number") is a branch of mathematics and the science of natural, real, concrete numbers—having a material existence, denoting a thing, not a quality or state—and zero under the operations of addition, subtraction, multiplication, and division. It involves **whole numbers** that are used to record the solutions of many arithmetic problems that can be solved by counting or by grouping objects and **fractions** to record the solutions of other problems solved by measuring and comparing quantities. When comparing two quantities, there is a **ratio,** a measurable relation of one thing to another. The ratio of two whole numbers is called **fraction**. A fraction *simply* means a fracture—the part being broken.

Arithmetic is the first branch of mathematics that depicts the science of numbers and the art of reckoning by figures or symbols.

It encompasses the representation of identification marks by symbols called **rational** numbers, whole numbers and fractions, and **irrational numbers** such as $\sqrt{2}$, $\sqrt{3}$, e, π pi, ø phi, etc.

Arithmetic Fundamental Theorem

Every *natural number* (except 1) can be written as a *product of prime numbers* in exactly one way (ignoring rearrangements). *Prime numbers* are the irreducible building blocks of *all natural numbers.*

Whole Numbers

The decimal numeral system is compact and allows a person to make calculations rapidly with pencil and paper. It is based on ten digits or numbers: 0, 1, 2, 3, 4, 5, 6, 7, 8, 9; and its most important feature is the place value; for example, 2 means two ones; 20 means two tens and no ones; and 200 means two hundreds, no ones, and no tens.

Addition is a way of putting together two or more similar things to find the sum. **Subtraction** is the opposite of addition. **Multiplication** with whole numbers is a short way of **addition**. **Division** is a way to separate things into equal parts.

The multiplication process involves the following: **multiplicand** (a number to be multiplied by another number), **multiplier** (number that multiplies the multiplicand) and **product** (the number obtained by multiplying two or more numbers together). For example:

$4 \times 2 = 8$ 4 is the multiplicand, 2 is the multiplier, and 8 is the product.

The division process involves the following: division are the **divisor** (number that divides the dividend), **dividend** (number to be divided by another number), **quotient** (number of times one quantity contains in another), and **remainder**. The following is an example with no remainder:

$$4\overline{)8}^{\,2} = 4$$ is the *divisor*, 8 is the *dividend,* and 2 is the *quotient.*

Fractions Are Divisions

As it was stated before fractions are divisions. For example, if we have a whole cake, we can divided into fractions:

Mary has taken **6** pieces of cake, and Joe had **8** pieces of cake. The **ratio** between Mary's pieces to Joe's pieces is 6 to 8, or 6 ÷ 8.

Written as a **fraction:** $\frac{6}{8}$ *

Written in **decimal form:** 0.75

Written in **percentage form:** 75%

* To convert fraction into decimal, just make a simple division—6 ÷ 8 = 0.75.

Fraction $\frac{6}{8}$ ← $\frac{\text{Numerator}}{\text{Denominator}}$ * = **Division** $\frac{6}{8}$ ← $\frac{\text{Dividend}}{\text{Divisor}}$ = 0.75 or 75%

* Tip to remember which is the denominator—remember d as **down.**

If the whole cake has 16 pieces, the fraction $\frac{6}{16}$ shows that Mary took 6 out of the 16 slices.

Percentage (%) is another way of expressing parts of a whole (100%). It can be written as a number, fraction, or decimal expressed in hundredths. Some of the decimals that go on forever are repeated

The cycle that repeats is usually denoted with a bar over the numbers. For example,

$$0.33333\ldots = 0.\overline{3}$$

Mixed Numbers

Convert all *mixed numbers* to *improper fractions* before adding, subtracting, multiplying, or dividing and then convert them back to a mixed number for your final answer.

Example: $10\dfrac{1}{4} - 5\dfrac{1}{3} = \dfrac{41}{4} - \dfrac{16}{3} = \dfrac{41\times 3 - 16\times 4}{4\times 3} = \dfrac{59}{12} = 4\dfrac{11}{12}$

<table>
<tr><td>Arithmetical
Progression</td><td>Geometrical
Progression</td></tr>
</table>

$$1 + 2 + \ldots + n = \frac{1}{2}n(n+1) \qquad 1 + a + a^2 + \ldots + (a-1)^{n-1} = \frac{1 - a^{n+1}}{1 - a}$$

Converting Percentages to Fractions and Decimals

The number of zeros in the denominator correspond to the number of digits after the decimal point in percentage form.

Number	Fraction	Decimal
175.00%	$\dfrac{175}{100}$	1.75
75.00%	$\dfrac{75}{100}$	0.75

55.50%	$\dfrac{55.5}{100}$	0.555
5.00%	$\dfrac{5}{100}$	0.05
1.50%	$\dfrac{1.5}{100}$	0.015

*   All fractions can be reduced. For example,

$$37.50\% = \frac{37.5}{100} = \frac{375}{1000} = \frac{125 \times 3}{125 \times 8} = \frac{3}{8}$$

$$5.00\% = \frac{5}{100} = \frac{5 \times 1}{5 \times 20} = \frac{1}{20}$$

$$66\frac{2}{3}\% = \frac{66\frac{2}{3}}{100} = \frac{\frac{66 \times 3 + 2}{3}}{100} = \frac{3}{3} \times \frac{\frac{200}{3}}{100} = \frac{200}{300} = \frac{2}{3}$$

Percentage Problems

Example 1: Percent increase

Kathy knew 144 phrases one year ago; now she knows 12.5% more phrases. How many does she know now?

$$144 \times \left(1 + \frac{12.5}{100}\right) = 144 \times (1 + 0.125) = 144 \times 1.125 = 162$$

Answer: 162 phrases

Example 2: Percent decrease

The dining set's original price is $450. How much does it cost during the store's 35% sale?

$$450 \times \left(1 - \frac{35}{100}\right) = 450 \times (1 - 0.35) = 450 \times 0.65 = 292.5$$

Answer: $292.50

Example 3: Simple interest

$$I = \frac{P \times r \times t}{100} \; (Interest = Payment \text{ x interest } rate \text{ percentage x } time)$$

Rose invests $2,500 at 2.5% simple interest. How much money would she have in 8 months?

$$2{,}500 \text{ x } \frac{2.5}{100} \text{ x } \frac{8}{12} = (2{,}500 \text{ x } 0.025 = 0.062500 \text{ x } 0.666666 = 41.67)$$

Answer: $2,541.67

Example 4: Compound interest

If Rose's interest is compounded monthly, she would have:

$$2{,}500 \text{ x } \left(1 + \frac{2.5}{100 \text{x} 12}\right)^{\frac{8}{12} \text{x} 12} = 2{,}541.97 \text{ (difference of 30¢)}$$

Note: When dealing with percentage problems, always look for the word *of* between the 2 numbers, which means multiplication, and *is* and *is of* between the 2 numbers, which mean division. For example,

What **is** __% **of** __? What % **of** __ **is** __? __ is what % **of** __?

$$\underline{40} \qquad \underline{56}$$

$$= \frac{40}{100} \times 56 \qquad \frac{\underline{56}}{40} \quad \frac{\underline{40}}{56} \quad \underline{56} \qquad \qquad \underline{40}$$

$$= .40 \times 56 \quad = 56 \div 40 \qquad = 40 \div 56$$

$$= 22.4 \qquad \qquad = 0.71 \qquad \qquad = 1.4$$

__ **is** __% **of what number?**

$$\underline{20} \qquad \qquad \underline{8}$$

$$= \frac{20}{8} \times 100 \times 100$$

$$= 2.5 \times 100 = 250$$

Decimals are <u>fractions</u> expressed as part of the decimal number system. For example,

Place Value

4,863,179,328.75

To the left of decimal point:

8 indicates ones

2 indicates tens

3 indicates hundreds

9 indicates thousands

To the right of decimal point:

7 indicates tenths

5 indicates hundredths

7 indicates ten thousands

1 indicates hundred thousands

3 indicates millions

6 indicates ten millions

8 indicates hundred millions

4 indicates billions

Note: The number 328.75 means 3 hundreds, 2 tens, 8 ones, 7 tenths, and 5 hundredths; and it can also be written as $328 \frac{75}{100}$

Example: $\frac{4}{10} = 0.4$; $\frac{4}{10} = 0.4$; $\frac{4}{1000} = 0.004$

Reduction or Cancellation of Common Fractions

$\frac{3}{4} \ \frac{6}{8} \ \frac{9}{12} \ \frac{75}{100}$ All of these fractions have the same value.

To reduce these fractions to the same value, divide the numerator and the denominator by the **same number**.

For example, in the fraction $\frac{9}{12}$ divide the numerator by 3 and divide the denominator by 3 and get $\frac{3}{4}$

Note: Two fractions are equal if their **cross products are equal**. For example, using the first two previous fractions, $\frac{3}{4}$ and $\frac{6}{8}$, we have:

$3 \times 8 = 24$

$4 \times 6 = 24$

Proportion. Two equal fractions make a proportion. For example,

$$\frac{3}{4} = \frac{6}{8}$$

Least Common Denominator (LCD)

LCD is crucial to performing operations with fractions. It brings all fractions to the smallest and the same **denominator**. The most important component of the LCD is the **divisor of the denominator**.

To find the LCD of two fractions, the **divisors** of the second **denominator** are multiplied by the first **denominator**, or vice versa if that is the case. Finally, the LCD is found by dividing the products by the second denominator to assess if they can be divided, and the least common denominator is chosen.

The LCD, also expressed as *least common multiple* (LCM) of two numbers or fractions, is the **smallest** number that is divisible by both.

LCD of 2 Numbers

Factor Tree   **Factor Tree**

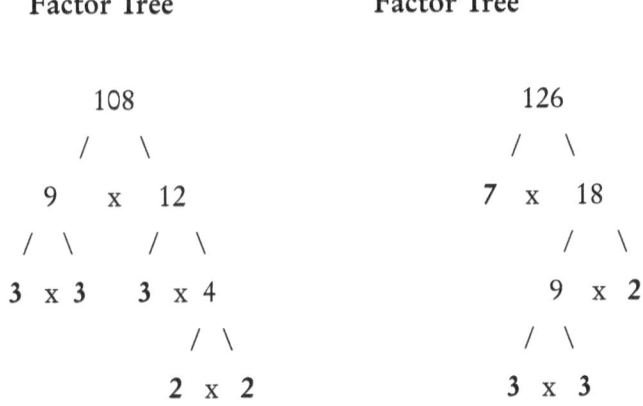

Thus, **prime factorization** of original number

$$108 = 2 \times 2 \times 3 \times 3 \times 3 = 2^2 \times 3^3.$$

prime factorization of original number

$$126 = 2 \times 3 \times 3 \times 7 = 2 \times 3^2 \times 7.$$

Two numbers Common Factors GCF LCD

108 and 126 $1, 2, 3, 6, 9,$ and $\mathbf{18}$ $\mathbf{18}$ $2 \times 2 \times 3 \times 3 \times 3 \times 7 = 2^2 \times 3^3 \times 7 = \mathbf{756}$

Note: GCF stands for greatest common factor

$$\text{Since the } GCF \text{ is } \mathbf{18}, \text{ their LCD} = \frac{108 \times 126}{\mathbf{18}} = \mathbf{756}$$

Note: When there is a need to find the *LCD* of *composite numbers* with no *common factors other than* **1** (GCF is **1**)—they are called "relatively prime"—we just multiply them and their *LCD* is their *product.*

Example: The *LCD* of 21 and 40 is **840** : 21 40 = **840**

LCD of Two Fractions

First example: Subtraction of Fractions

$$\frac{3}{4} - \frac{1}{6}$$

The **divisors** of 6 are 1, 2, 3, and 6.

Multiply and Divide

$1 \times 4 = 4$ No.   4 cannot be evenly divided by 6

$2 \times 4 = 8$ No.   8 cannot be evenly divided by 6

$3 \times 4 = 12$ Yes. 12 can be divided evenly by 6

$6 \times 4 = 24$ No. Even though 24 can be divided evenly by 6, it is larger than 12

So the **LCD** is 12.

Now combine the original denominators and numerators.

Divide and Multiply

$12 \div 4$ (denominator) $= 3$ Then 3×3 (numerator) $= 9$

$12 \div 6$ (denominator) $= 2$ Then 2×1 (numerator) $= 2$

The result is: $\dfrac{9}{12} - \dfrac{2}{12} = \dfrac{7}{12}$ **The final answer is:** $\dfrac{3}{2} - \dfrac{1}{6} = \dfrac{7}{12}$

To find the LCD of three or more numbers, multiply the first denominator by the second denominator; then multiply the second denominator by the third denominator; and lastly, multiply the first denominator by the third denominator. Once you have the three products, determine which is the LCD for all three denominators.

Second Example: Mixed Operations

$$\frac{5}{8} + \frac{4}{6} - \frac{2}{3}$$

Multiply and Divide

$8 \times 6 = 48$ **No.** Even though, **48** can be divided evenly by **8, 6, 3**

$6 \times 3 = 18$ **No. 18** cannot be divided evenly by **8**

$8 \times 3 = 24$ **Yes. 24 is the smallest** number that can be divided evenly by **8, 6, 3**. So **24** is the **LCD**.

Now combine the original denominators and numerators.

Divide and Multiply

Denominator **8** divided by **24** = **3**. Then, **3** multiplied by numerator **5** = **15**

Denominator **6** divided by **24** = **4**. Then, **4** multiplied by numerator **4** = **16**

Denominator **3** divided by **24** = **8**. Then, **8** multiplied by numerator **2** = **16**

The result is: $\dfrac{15}{24} + \dfrac{16}{24} - \dfrac{16}{24} = \dfrac{15+16 = 31-16}{24} = \dfrac{15}{24} = \dfrac{5}{8}$

Final Answer is: $\dfrac{5}{8} + \dfrac{4}{6} - \dfrac{2}{3} = \dfrac{5}{8}$

Third example: Operations involving whole numbers and fractions

$$2 - \dfrac{1}{3} - \dfrac{1}{4} = 2 - \dfrac{4}{12} - \dfrac{3}{12} = 24 - \dfrac{4}{12} - \dfrac{3}{12} = \dfrac{17}{12}$$

Comparison of Fractions

Which is larger? $\dfrac{3}{5}$ or $\dfrac{7}{12}$

Taking into consideration that $\frac{5}{5} = 1$ and $\frac{12}{12} = 1$, the answer is $\frac{12}{12}$

Notice that the first fraction just needs two digits in its numerator to make a whole number, and the second fraction needs 5 digits in its numerator to make a whole number.

This comparison can be resolved by deduction or by the following operation:

1) Find the common denominator: Multiply 5 by 12 = 60
2) Multiply 3 by 12 = 36 and then multiply 7 by 5 = 35
3) The results are:

$$\frac{3}{5}, \frac{7}{12} = \frac{36}{60}, \frac{35}{60} \quad \text{Answer:} \quad \frac{36}{60} \text{ or } \frac{3}{5}$$

Adding, Subtracting, Multiplying, and Dividing Fractions

All of these four basic operations to solve problems with fractions must be performed once we have simple common fractions, and the common denominator is needed **only** in **addition** and **subtraction**.

Simplify: $8\frac{2}{3} = \frac{26}{3}$ To convert a mixed number into a simple fraction, just multiply the whole number by the denominator then add the numerator. Place the result as the new numerator and keep the previous denominator. For example,

$$2\frac{3}{4} = \frac{11^*}{4} \qquad 15\frac{1}{2} = \frac{31}{2}$$

* $\dfrac{11}{4}$ can also be represented as $\dfrac{8}{4} + \dfrac{3}{4}$.

Remember that fractions are divisions, thus

The numerator is the dividend,

The denominator is the divisor.

Improper Fractions (numerator is larger than denominator) *

$$\frac{8}{1}=8; \quad \frac{8}{2}=4; \quad \frac{8}{3}=2.6; \quad \frac{8}{4}=2; \quad \frac{8}{5}=1.6; \quad \frac{8}{6}=1.3; \quad \frac{8}{7}=1.1; \quad \frac{8}{8}=1$$

* $8 = \dfrac{8}{1}$ both represent a whole number.

Proper Fractions (numerator is smaller than denominator)

$$\frac{1}{8}=0.1; \quad \frac{2}{8}=0.2; \quad \frac{3}{8}=0.4; \quad \frac{4}{8}=0.5; \quad \frac{5}{8}=0.6; \quad \frac{6}{8}=0.7; \quad \frac{7}{8}=0.9; \quad \frac{8}{8}=1$$

Note: $\dfrac{0}{8}=0$; $\dfrac{0}{8}$ is **undefined; and the reciprocal of a fraction is its inverse. For example,**

The reciprocal of $8 = \dfrac{8}{1}$ is $\dfrac{1}{8}$; and $x - 2 = \dfrac{x-2}{1}$ is $\dfrac{1}{x-2}$

Note: Don't confuse **exponent -1 in** x^{-1}, with the **superscript -1 in** f^{-1}.

x^{-1} is the *reciprocal* of something — that is, $x^{-1} = \dfrac{1^*}{x}$

$f^{-1}(x)$ is the *inverse* of $f(x)$ — that is, $f(x) = \sqrt{x}$

*   It means that negative power -1 is converted in positive power 1 (power 1 is *always* omitted).

Need common denominator

Adding Fractions   Example: $\dfrac{2}{7} + \dfrac{4}{7} = \dfrac{6}{7}$

Subtracting Fractions   Example: $\dfrac{4}{7} - \dfrac{2}{7} = \dfrac{2}{7}$

Do not need common denominator

Multiplying Fractions   Example: $\dfrac{2}{5} \times \dfrac{4}{4} = \dfrac{8}{20} = \dfrac{4}{10} = \dfrac{2}{5}$

Dividing Fractions*   Example: $\dfrac{3}{10} \div \dfrac{4}{5} = \dfrac{3}{10} \times \dfrac{5}{4} = \dfrac{15}{40} = \dfrac{3}{8}$

*   Dividing is almost the same as multiplying. Just flip the second fraction, and then proceed to multiply. Both do not need the same denominator; but when adding and subtracting, a *common denominator* is needed.

All fractions need to be reduced to the *simplest form* For example,

$\dfrac{8}{20} = \dfrac{4}{10} = \dfrac{2}{5}$ or when splitting a fraction: $\dfrac{5}{6} = \dfrac{3}{6} + \dfrac{2}{6} = \dfrac{1}{2} + \dfrac{1}{3} = \dfrac{5}{6}$

$(\dfrac{1}{2} + \dfrac{1}{3},$ *find LCD* 2 x 3 = **6,** *cross multiply* $1 \times 3 = $ **3** and 2 ×1 = **2,** and *substitute*):

Therefore,

$$\dfrac{5}{6} = \dfrac{1}{2} + \dfrac{1}{3} = \dfrac{3}{6} + \dfrac{2}{6} = \dfrac{5}{6}$$

$$\dfrac{17}{20} = \dfrac{1}{4} + \dfrac{3}{5} = \dfrac{5}{20} + \dfrac{12}{20} = \dfrac{17}{20}$$

Absolute value

Absolute value just turns a negative number into a positive number and does not affect a positive number or zero. For example,

$$| -3| = 3, \ |3| = 3, \ |0| = 0$$

Absolute value of a number is its distance from zero in a number line.

Powers

Any number—a fraction, a negative number, etc.—raised to the zero power equals 1. Zero raised to the zero power is undefined. If the bases of two powers are the *same*, it is possible to *add or subtract their exponents*. For example,

Add exponents Subtract exponents

$$2^3 \times 2^8 = 2^{11} \quad 2^8 \div 2^3 = 2^5$$

Powers are defined in the following manner:

Zero **power**   For example, $x^0 = 1$

One **power**   For example, $12x^1 = 12$ *

---

*   Any number to the first power is *itself*. An exponent of 1 is always omitted, thus, $12x^1$ is written $12x$ only.

Second **power** or squared   For example, $0^2 = 0$

$$1^2 = 1$$

$$2^2 = 4$$

$$3^2 {}_* = 9$$

*   It is read as: **3** is raised to the second power or squared: 3 x 3 = **9**

All of the above are *perfect squares*. Order of operation:

$$4^2 = 16$$

$$(-4^2) = 16$$

$$-4^2 = - (4^2) = -16$$

Third **power** or cubed For example, $2^3 {}_* - 8$

*   It is said that 2 is raised to the third power or cubed.

Fourth **power** and thereafter For example, $4^4 {}_* = 256$

*   It is said that 4 is raised to the fourth power.

Power Rules

Addition of powers is done *only* when terms are separated with a multiplication sign.

$$\text{Example: } x^2 \bullet x^3 = x^5$$

Subtraction of power is done *only* when terms are in a fraction form (i.e. when terms are being divided).

$$\text{Example: } \frac{x^5}{x^3} = x^2$$

Multiplication of powers *only* when term is enclosed in parenthesis.

$$\text{Example: } (x^2)^3 = x^6$$

Distribution of powers *only* when variables are enclosed in parenthesis or in a form of fraction <u>without any minus or plus signs.</u>

$$\text{Example: } (xyz)^3 = x^3 y^3 z^3$$

$$\frac{(x)^4}{y} = \frac{x^4}{y^4}$$

Factor powers *only* when <u>there are plus or minus signs</u> between variables.

$$(x + y)^2 = (x + y)\,(x + y) \qquad (x\text{-}y)^2 = (x - y)\,(x - y)$$

$$= x^2 + xy + yx + y^2 \qquad\quad = x^2 - xy - yx - y^2$$

$$= x^2 + 2\,xy + y^2 \qquad\qquad = x^2 - 2yx - y^2$$

Negative and Fraction Powers

These powers are tricky, so it is important to know how they function.

Examples: $x^{-3} = \dfrac{1}{x^3}$ $5^{-1} = \dfrac{1}{5^2} = \dfrac{1}{25}$

$x^{-2} = \dfrac{1}{x^2}$ **or** $x^{-2} = -2x^{-3}$

$x^{-2/3} = \dfrac{2}{3}x^{1/3}$ $\dfrac{5}{4}x^{1/3} = \dfrac{1(5)}{3(4)}x^{-2/3} = \dfrac{5}{12}x^{-2/3}$

Note: A negative number raised to even power is a **positive number.**
A **positive number** raised to any power is a **positive number.**

Examples:

$$(-3)^4 = 81 \text{ and } (3)^4 = 81$$

$$(-1)^4 = 1 \text{ and } (1)^4 = 1$$

$$60(-1)^4 = 60 \text{ and } 60(1)^4 = 60$$

$$15(-1)^4 = 15 \text{ and } 15(1)^4 = 15$$

$$60(3)^2 = 60 \cdot 9 = 540 \text{ and } 60(-3)^2 = 60 \cdot 9 = 540$$

Power rules

1) **Convert negative powers into positive powers:** $4^{-2} = \dfrac{1}{4^2} = \dfrac{1}{16}$

This results is $\dfrac{1}{16}$, which is not negative.

2) Convert a root problem into an easier power problem.

Any root problem can be converted into a power problem. For example,

$$\sqrt[3]{x} = x^{1/3}; \quad \sqrt{x} = x^{1/2}; \quad \sqrt[4]{x}^{3} = x^{3/4}$$

*   Remember that x = 1

3) Addition, Subtraction, and Multiplication of Powers

Powers are added with the sign of multiplication; powers are subtracted when they are used in fractions; powers are multiplied when they are enclosed in brackets.

Square Root

A square root is a **number** that when multiplied by itself equals a given number. The symbol for a square root is called a radical sign. For example:

$$\sqrt{25} = 5, \text{ because } 5 \times 5 = 25; \quad \sqrt{9} = 3; \quad \sqrt{4} = 2, \text{ etc.}$$

Some are not so obvious:

$$\sqrt{2} = 1.41 \text{ and } \sqrt{3} = 1.73$$

An easy way to find roots is by using a table of square roots or a slide rule.

Simplifying Roots

There are two ways to simplify roots:

1) Example: $\sqrt{300} = 10\sqrt{3}$ because 300 equals 100 times 3.

2) Example: $\sqrt{504}$

   - Break down 504 into a product of all its **factors** $\sqrt{2.2.2.3.3.7}$
   - Enclose each pair of number [22] and [33]
   - For each enclosed pair take one out: $2.3\sqrt{2.7}$
   - Simplify: $2.3\sqrt{2.7}$ and this is the **answer**.

3) Examples:

$$\sqrt{0} = 0; \quad \sqrt{1} = 1; \quad \sqrt{2} = 1.41; \quad \sqrt{3} = 1.73; \quad \sqrt{4} = 2; \quad \sqrt{5} = 2.24;$$
$$\sqrt{6} = 2.45; \quad \sqrt{7} = 2.64; \quad \sqrt{8} = 2.83 \quad \sqrt{9} = 3; \quad \sqrt{10} = 3.1623$$

4) Examples of roots and its equivalent power: $\sqrt{x} = x^{1/2}$

$$3\sqrt{x} = x^{1/3}; \quad \sqrt[3]{x^2} = x^{2/3}; \quad \sqrt[4]{x} = x^{1/4}; \quad \sqrt[4]{x^2} = x^{2/4}; \quad \sqrt[4]{x^3} = x^{3/4}$$

5) Example of a square root of product:

$$\sqrt{9 \times 4} = \sqrt{9} \times \sqrt{4} = 3 \times 2 = 6$$

6) Example of a square root of quotient:

$$\sqrt{\frac{64}{4}} = \frac{\sqrt{64}}{\sqrt{4}} = \frac{8}{2} = 4$$

7) Example of a *simplified* square root expression. Radicand has no repeated factors because they are removed outside the square root:

$$\sqrt{60} = \sqrt{2 \times 2 \times 3 \times 5} = \sqrt{2 \times 2}\,\sqrt{3 \times 5} = {}^2\sqrt{15}$$

8) Example of a cube root:

$$\sqrt[3]{512} = 8$$

Logarithms

A logarithm is just a different way to represent an exponential expression.

Exponential expression $2^3 = 8$ is represented as a logarithm:

$Log_2 8 = 3$, it reads as "log base 2 of 8 equals 3.

The base of logarithm can be any number greater than zero and other than one; if the base is 10 it is not written. Log base e (e ≈ 2.72) is written *ln* instead of log.

Logarithm Properties

$$a^{\log_a b} = b$$

$$\log_a a^b = b$$

$$\log_a b = \frac{\log_c b}{\log_c a}$$

$$\log_c a^b = b\log_c a$$

$$\log_c (ab) = \log_c a + \log c\, b$$

$$\log_c \frac{(a)}{b} = \log_c a - \log c\, b$$

$$\log_c c = 1$$

$$\log_c 1 = 0$$

Chapter 3

DISCLOSING ABSTRACT ALGEBRA

A lgebra is a branch of mathematics that uses sets of **abstract numbers or symbols** on which certain formal operations are defined; For example, (**x, y**) represents x and y coordinates on the x-y coordinate system, and **a, b, c** represents a notation that must be consistent and unambiguous, the same symbol should denote the same thing whenever it occurs. The basic algebraic system is the **group** in which there is **one operation**. Rings and fields have both multiplication and addition and more closely resemble ordinary concrete number system; the **real** and complex number systems are fields. Algebra is based on two important concepts: **vector space** and **functions.**

Fundamental Theorem of Algebra

Every algebraic equation has a solution *if* it is interpreted on the terms of the *system of complex numbers.*

$$\text{Pair number} \quad\quad \text{x, y}$$
$$= (0, 1) = i$$
$$= (0, -1) = -i$$

Note: x, y = x + dy **and** x,y = x + iy

*   Square root of negative one: *Denote and Replace:* $\sqrt{-1}$ **by** i, and i^2 **by** $-i$. Number (-1) *has no square root* and number pair (0, -1) has two square roots.

What is a **vector space?**

Vector space is an abstract system with formal properties similar to those used in ordinary vector algebra. **Vector,** or vehicle, is a straight line of definite length drawn from a given point in a given direction, representing a quantity or parameter (as a velocity or a force), which has both magnitude and direction.

What are **functions?**

Functions are performances/relationships within two **variables, x** and **y**, so connected with each other that any change in one affects and produces a corresponding change in the other. **Functions,** or mappings from one algebraic to another, preserve the operations of the system and play an essential role in abstract algebra. *In mathematics function is written f(x).*

What are **variables?**

A **variable** is a quantity that may have any one set of values; (x,y) is called complex number or pair number. Complex variables can be (x,y) or x, y, i, or z. *In algebra x is an **unknown variable**.*

Functions or mappings of vector spaces can be represented by **matrices, X,** which have also been an important part in abstract algebra. Matrices are rectangular arrays of quantities or symbols and are also sources in which everything is formed—that in which anything is embedded.

What is a limit?

A *limit* is a process of getting a **finite number** through both a *derivative* and an *integral.*

What are *representations of a* function, tangent line, derivative, normal line, secant line, slope, and points?

Function = Vertical line (y) that touches the *curve at only one point.*

Tangent line = Line that intersects a *curve* at *only one point.* The **slope** of the tangent line is the **derivative.**

Derivative = The most fundamental meaning of a derivative is that it is a **rate** like miles **per** hour (miles as a *function* of hours). The derivative becomes the **slope** of the *function.* Thus, the *derivative* is a *rate* which on a graph appears as a *slope.*

Normal line = Line that is drawn perpendicularly to a tangent line at the point of tangency.

Secant Line = Line that intersects a *curve* at *two points.*

Slope = Steepness of a *curve* that is determined by taking *two points on the curve* and plugging them into the formula

$$\frac{y2-y1}{x2-x1}$$

For example, 2 points (6,36) and (2,4)
$$= \frac{36-4}{6-2}$$
$$= \frac{32}{4}$$

Slope   **= 8**

Point = (x,y) is represented by 2 variables -x,y coordinates; For example, (6,36) is represented by 2 numbers -6 comes from the **horizontal x line-coordinate** and 36 from the **vertical y-line coordinate.**

What is a **term?**

A term is an **algebraic expression** that represents each quantity in a series.

In $4x^5$, 4 is a **coefficient,** x is a **variable,** and 5 is an **exponent or power.**

$$4x^5$$

A **coefficient** is the numeral factor prefixed to an unknown quantity—the **variable,** and the **power** represents the number of times that a given number occurs as a factor in a product.

What is an **algebraic expression?**

An **algebraic expression** is composed of a term or several terms, and each algebraic expression is categorized according to the number of terms.

Single Term: $8x^3$

Binomial—Two Terms: $8x^3 + y^2$

Trinomial—Three Terms: $8x^3 + y^2 + 12x^5$

Polynomial—More than 3 terms: $8x^3 + y^2 + 12x^5 5x + 2$

Note: Each term is separated by a sign, (+) or (-), as in the above examples.

What is an **equation?**

An equation is when the two sides are *equivalent* to each other.

What is a **quadratic equation?**

A **quadratic equation** is composed of terms whose greatest exponent is 2.

What is a **polynomial?**

Representation/description of a Polynomial (four terms)

$$4x^3 2y^4 + 6x^2 y^5 - 20x^4 y^3 z^2 - 5x + 2$$

A polynomial is an expression as the one described above in which all of its terms have a variable raised—to a positive integral power, except for the **constant 2**. A constant is a single number with no variable or power. In other words, fraction or negative powers, radicals, sines or cosines, etc., are not allowed; just terms with a coefficient multiplied by a variable raised to a power. The **degree** of a polynomial is the highest power of any of the variables; in the case above, the polynomial has a degree of 5, and it is the variable y that has the power of 5. First degree polynomial is one like **x**, which has a power of 1 that is *never written.*

Factoring—A Powerful Tool

What is **factoring?**

Factoring is **simply** done to get the factors derived from one number. **Factors** are integers (natural numbers) which can be divided into another integer without any remainder; For example, **1, 2, 3, 4,** and **6** are factors of **12** since each of these may be used to divide 12 into exact integers. Factors are also numbers that when multiplied by another number equal a given number; For example, **4 x 3 = 12; 6 x 2 = 12;** and **12 x 1 = 12.** Every *natural number greater than* 1 can be factored into a product of primes. The method for factorization is called *factor tree,* and the end result is called *prime factorization* of the original number.

In calculus, there is always a need to **factor** algebraic expressions. The goal in algebraic expressions is to find the **GCF** (greatest common **factor**). Algebraic factoring always involves rewriting a *sum of terms as a product.* Therefore, it is important to see the similarity of finding the GCF with natural numbers and with algebraic expressions to dissipate its mystery. Once there is clarity, it is obvious that factoring means "unmultiplying," in other words, like rewriting.

Natural number	24
as:	2 x 2 x 2 x 3
Algebraic expression	5xy + 10yzxx
as:	5y (x + 2z)

*   Notice that the parenthesis is used to signify multip–lication.

The GCF which can also be expressed as GCD (greatest common divisor) of two numbers *is* the **largest** of the *common factors*. And the *common factor* of two numbers is **any number that is a factor of both,** For example,

Two numbers	*Common factors*	*GCF*
108 and 206	2, 3, 9, **6, 18**	18
		(**2 x 3 x 3**)

```
        Factor Tree          Factor Tree
           108                  126
          /  \                 /  \
        9  x  12             7  x  18
       / \  / \                  / \
     3 x 3 2 x 6              2  x  9
            / \                  / \
          2 x 3                3  x  3
```

Thus, **prime factorization** of original number

$$108 = 2 \times 2 \times 3 \times 3 \times 3 = 2 \times 3$$

prime factorization of original number

$$126 = 2 \times 3 \times 3 \times 7 = 2 \times 3^2 \times 7$$

What is a **linear factor?**

A *linear factor* is any *first degree* polynomial (first degree is power of 1 that is not written) as in the case of the following factored denominator that contains two linear factors with a power of 1:

$$\frac{5}{(x-2)(x+3)}$$

Dissipating the Mystery with Examples

Find the GCF for:

Natural numbers	Algebraic expression
$8 + 12 + 20$	$8x^3 y^4 + 12x^2 y^5 + 20x^4 y^3 z$
$\underline{GCF}\ 4\ (\ 2 + 3 + 5\ )$	$\underline{GCF}\ 4x^2 y^3\ (\ 2xy + 3y^2 + 5x^2 z\ )$

Note that in both operations, the same natural number for GCF is **4**. The difference resides in displaying the variables and powers in the second operation. Here you **subtract** the powers. Note that the variable exponent is not indicated when the exponent is 1.

Solving Quadratic Equations

There are three methods to solve quadratic equations:

Method 1: Finding GCF, looking for patterns, and factoring

Method 2: Quadratic formula

Method 3: Completing the square

Method 1: Factoring

First Step: The first step in factoring any type of expression is to pull out or factor out the **GCF**.

Second Step: The second step is to look for one of three **patterns**:

Difference of squares: $a^2 - b^2 = (a - b)(a + b)^*$

Difference of cubes: $a^3 - b^3 = (a - b)(a^2 + ab + b^2)$

Sum of cubes: $a^3 + b^3 = (a + b)(a^2 - ab + b)^2$

* a - b = (a - b)(a + b) is the same as

$$(this)^2 - (that)^2 = (this - that)(this + that)$$

So rewrite $9x^2$ -25 to look like the **pattern:** $9x^2 - (5)^2 = (9x -5)$ (9x + 5)

Note: Sum of squares cannot be factored. Be careful with the *signs* when factoring; For example,

$$a^2 - b^2 = (a - b)(a + b)$$
$$a^3 - b^3 = (a - b)(a^2 + ab + b^2)$$
$$x^2 + x - 6 = (x - 2)(x + 3)$$

(+ - = -)

$$A^3 + b^3 = (a + b)(a^2 - ab + b^2)$$

$$2x - x - 12 = (2x + 3)(x - 4)$$

$$(x^2 - y) = (x + y)(x - y) = x^2 - 2xy - y^2$$

(- - or ++ = +)

$$(x + y)^2 = (x + y)(x + y) = x^2 + 2xy + y^2$$

Adding, Subtracting, Multiplying, and Dividing *Signed Numbers*

Signs Rules

$$+ + = +$$
$$- - = +$$
$$+ - = -$$
$$- + = -$$

When **adding,** just take the sign of the *bigger number* and use it for your final answer. When **subtracting,** switch the sign of the *bigger number* and use it for you final answer.

Additions

Examples:

$$-5 + (-1) \quad = \quad -(5+1) \quad = \quad -6$$
$$-6 + 14 \quad = \quad +(14-6) \quad = \quad 8$$
$$7 + (-11) \quad = \quad -(11-7) \quad = \quad -4$$

Subtractions

Examples:

$$-5 - (-14) \quad = \quad -5 + 14 \quad = \quad 9$$
$$3 - (5-1) \quad = \quad 3 + (-5) + 1 \quad = \quad -1$$
$$4 - (3 - 5 \times (3)) \quad = 4 - 3 + 5 \times (-3) \quad = \quad -14^{*}$$

*   Flip *only once* for a product of two numbers.

Multiplications and Divisions

When multiplying or dividing numbers, just follow the *Signs Rules.*

Examples:

$$(-4) \times (-2) \quad = \quad 8$$
$$6 \times 3 \quad = \quad 18$$
$$(-3) \times 4 \quad = \quad -12$$
$$20 \times (-3) \quad = \quad -60$$
$$(-20) \div (-4) \quad = \quad 5$$
$$4 \div (-3) \quad = \quad -\frac{4}{3}$$

*   Remember, a division is just a fraction and vice versa. The denominator is the divisor and the numerator is the dividend.

Example 1:

Solve the quadratic equation: $2x^2 - 5x = 12$

1. Bring all terms on one side. Leave 0 on other side: $2x - 5x - 12 = 0$*
2. Factor and check with FOIL* $(2x + 3)(x - 4) = 0$
3. Set each factor equal to zero and solve the equation:

$$2x + 3 = 0 \qquad\qquad\qquad x - 4 = 0$$
$$2x = -3$$

Answer has two solutions:

$$x = -\frac{3}{2} \qquad \text{or} \qquad x = 4$$

$$x = -\frac{3}{2} \qquad \text{or} \qquad x = 4$$

*   FOIL stands for First, Opposite, Interior, and Last. In this example, you multiply the following:

First $2x$ and x = $2x^2$

Opposite $+2x$ and -4 = $-8x$*

Interior $+3$ and $+x$ = $\dfrac{+3x}{-5x}$ $-5x$

Last $+3$ and -4 = -12

_____

                            $2x^2$ $-5x$ -12

*   + + = positive $-$ $-$ = positive + $-$ = negative $-$ + = negative

Method 2: Quadratic Formula

The quadratic formula is probably one of the easiest methods to use in solving quadratic equations. The solution or solutions of a quadratic equation, represented as 2

$$ax + bx + c$$

are given in the quadratic formula.

Quadratic FORMULA

$$x = \frac{-b \pm \sqrt{b^2 - 4ac}}{2a}$$

Example 2:

Solve the quadratic equation $2x^2 - 5x = 12$

First Step: Bring terms on one side, leave 0 on other side $2x^2 - 5x - 12 = 0$

Second Step: Plug the coefficients into the formula.

In this example, $a = 2$; $b = -5$; and $c = -12$.

$$= \frac{-(-5) \pm \sqrt{(-5)^2 - 4(2)(-12)}}{2.2}$$

$$= \frac{-5 \pm \sqrt{25 - (-96)}}{4}$$

$$= \frac{5 \pm \sqrt{121}}{4}$$

$$= \frac{5 \pm 11}{}$$

$$= \frac{16}{4} \quad \text{or} \quad \frac{6}{4}$$

$$= 4 \quad \text{or} \quad \frac{3}{2}$$

Method 3: Completing the Square

This method creates a perfect square trinomial that can be solved by taking its square root.

Solve the quadratic equation: $3x = 24x + 27$

1. Place the **x** and **x** on one side. **Constant** on the other. $3x^2 - 24x = 27$

2. Divide both sides by the coefficient 3. $x^2 - 8x = 9$

3. Take half of the coefficient 8, square it, add both sides. $x^2 - 8x + 16 = 9 + 16$

4. Factor the left side. $(x - 4)^2 = 25$

5. Take square root of both sides, then place ± sign: $\sqrt{(x-4)^2} = \sqrt{25}$

$$x - 4 = \pm 5$$

6. Answer: $x = 4 \pm 5$

                  $= 9$ or -1

3 Important Definitions
(Function, Limit, and Derivative)

Function

A function is a relationship between two **variables (x,y)**. An x-y coordinate system is a simple graph with two lines—one **horizontal** representing **x** and one **vertical** representing y. **Note that y** and f(x) are exactly the same, and it *always* represents the different functions as

parabolic function presented as: $y = x^2$ or $f(x) = x^2$

Virtually all functions are represented by graphs (diagrams depicting the successive changes of a variable), which represent the lines or curves in this x-y coordinate system. The term *curve* refers to any shape, whether it is a line curved or straight. A **curve** is literally a function when it touches the **y vertical line only once**. This guarantees that each input has exactly one output. The curve is not a function if it touches the **y vertical line more than once**.

Function Curves

There are different types of functions, but the basic ones are the following:

Line	$y = 3x + 5$
Parabola	$f(x) = x^2$
Absolute Value	$f(x) = |x|$
Exponential Value	$f(x) = 2^x$
Logarithmic	$f(x) = \log_2$

Line Function is important in the study of calculus because we can determine its **derivative/rate**.

Limit

A limit is a process of getting a *finite number* through derivatives or integrals.

Derivative

A derivative comes from a Latin word *derivare*—*de* means "from," *rivus*, "river." *Derivative* is an adjective that means that it is derived or taken from a source or origin, and it is also *a limit of a sequence of ratios.* A derivative becomes a **rate** when it measures how much one thing changes compared with another. In order to visualize this **derivative** or **rate,** the graphs of the **curves/functions** on the x-y **coordinate system** are crucial. With these graphs, there is a clear picture of the curves/functions' **slope or steepness.** A good way to see the slope is to draw a stairway under the **curve**/line.

There are three conditions where the **derivative** does not exist.

1. There is no **tangent line,** and thus no derivative at any type of discontinuity, infinite, removable, or jump.
2. There is no **tangent line,** and thus no derivative at a cusp on a function.
3. When a function has a **vertical tangent line,** the slope is undefined; and therefore, the derivative fails to exist.

X-Y Coordinate System

Curve/function:

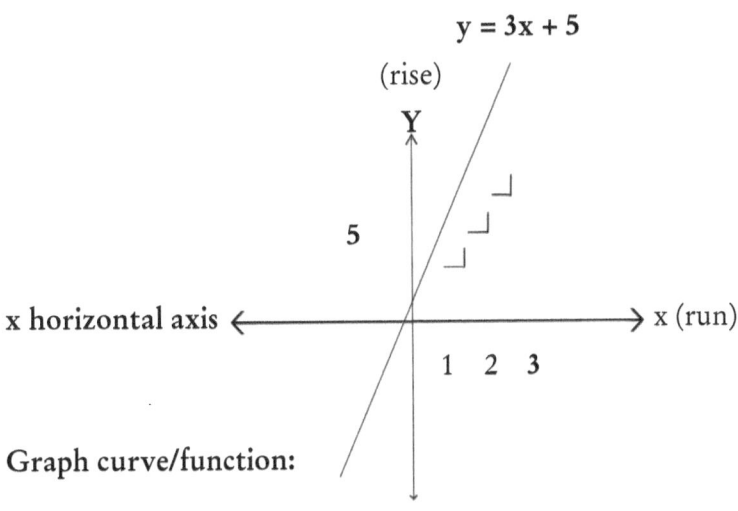

y = 3x + 5

(rise)

5

x horizontal axis ← ——————————— → x (run)

1 2 3

Graph curve/function:

y vertical line

Note: Each step of the above stairway is horizontally **1** unit in its width and vertically **3 units** in its length. Therefore, 3x is located at the end of the third step, thus **3x (run)** and **5** are located in the beginning of the first step, **5y (rise)**. This stairway begins at **5** in the y vertical line and ends at **3** in the x horizontal line. The vertical part of the step is called **run**, and the horizontal part of the step is called **rise**.

slope: $\dfrac{\text{rise}}{\text{run}} = \dfrac{3}{1} = 3$

Note: Because both the slope of 3 and the y-intercept of 5 appear in the equation y = 3x + 5, this equation is called *slope-intercept form.* In general,

$$y = mx + b$$

Lines that go up to the right have a *positive slope*; when they go down, they have a *negative slope*. Horizontal lines have a *slope of zero*, and vertical lines have an *undefined slope*.

Note: Exponential value and **logarithmic** means exactly the same, just written in different forms—the exponent or the power to which a fixed number has to be raised to produce a given number. In both cases, the **x** or **number** representing the power is written above and to the right of the algebraic expression. It is also important not to confuse the **exponent -1,** reciprocal of something, with the **superscript -1,** inverse of $f(x)$.

Trigonometry

Trigonometry is simply the study of the **right triangle.** The three main **trig functions (sine, cosine, and tangent)** and their reciprocals (**cosecant, secant, and cotangent**) deal with the lengths of the sides of a right triangle that contains an **acute angle**.

Pythagorean Theorem

$a^2 + b^2 = c^2$, where a and b are the lengths of the triangle and c is the hypotenuse. The Pythagorean theorem means that the sum of the squares of the two sides is equal to the square of the hypotenuse

45°   -45°   -90°
Right Triangle

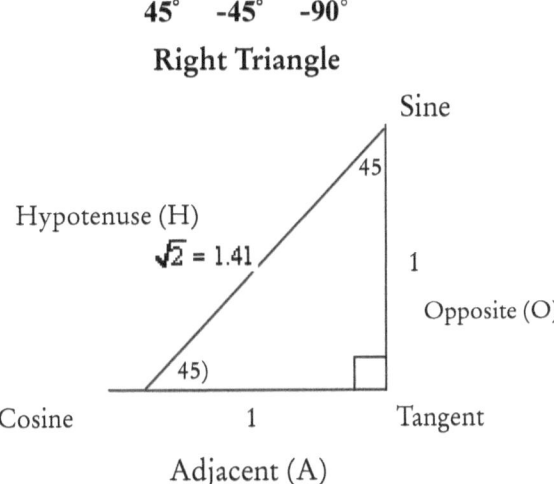

Note: To help remember the definitions of the trig functions, memorize the mnemonic *SohCahToa*, which uses the initial letters of *sine, cosine, tangent, hypotenuse, adjacent,* and *opposite.*

The Six Trigonometric Functions of an Angle

Sine	(sin)	is the ratio of the perpendicular (opposite side) to the hypotenuse identical with the cosine of the complementary angle.
Cosine	(cos)	is the ratio of the base (adjacent side) to the hypotenuse. identical with the sine of the complementary angle.
Tangent	(tan)	is the ratio of the sine to cosine, which is the same as the ratio of the perpendicular (opposite side) to the base (adjacent side).
Cosecant	(csc)	is the reciprocal of sine.

Secant   (sec)   is the reciprocal of cosine.

Cotangent   (cot)   is the reciprocal of tangent.

The trig values of the above triangle with angles 45°-45°-90°, which is half of a 1 × 1 **square** with a hypotenuse of $\sqrt{2} \approx 1.41$, are

$$\sin 45° = \frac{O}{H} = \frac{1}{\sqrt{2}} = \frac{\sqrt{2}}{2} \approx 0.71 \qquad \csc 45° = \frac{H}{O} = \frac{\sqrt{2}}{1} = \sqrt{2} \approx 1.4$$

$$\cos 45° = \frac{A}{A} = \frac{1}{\sqrt{2}} = \frac{\sqrt{2}}{2} \approx 0.71 \qquad \sec 45° = \frac{H}{A} = \frac{\sqrt{2}}{1} = \sqrt{2} \approx 1.4$$

$$\tan 45° = \frac{O}{A} = \frac{1}{1} = 1 \qquad\qquad \cot 45° = \frac{A}{O} = \frac{1}{1} = 1$$

The trig values of the right triangle with angles 30°-60°-90°, which is half of a 2 × 2 equilateral triangle with legs of lengths 1 and $\sqrt{3} \approx 1.73$ and a 2-unit long hypotenuse (remember hypotenuse is **always** the longest side of a right triangle), are

For the 30°:

$$\sin 30° = \frac{O}{H} = \frac{1}{2} \qquad\qquad\qquad \csc 30° = \frac{H}{O} = \frac{2}{1} = 2$$

$$\cos 30° = \frac{A}{H} = \frac{\sqrt{3}}{2} \approx 0.87 \qquad\qquad \sec 30° = \frac{H}{A} = \frac{2}{\sqrt{3}} = \frac{2\sqrt{3}}{3} \approx 1.15$$

$$\tan 30° = \frac{O}{A} = \frac{1}{\sqrt{3}} = \frac{\sqrt{3}}{3} \approx 0.58 \qquad \cot 30° = \frac{A}{O} = \frac{\sqrt{3}}{1} = \sqrt{3} \approx 1.73$$

For the 60°:

$$\sin 60° = \frac{O}{H} = \frac{\sqrt{3}}{2} \approx 0.87 \qquad \csc 60° = \frac{H}{O} = \frac{2}{\sqrt{3}} = \frac{2\sqrt{3}}{3} \approx 1.15$$

$$\cos 60° = \frac{A}{H} = \frac{1}{2} \qquad \sec 60° = \frac{H}{A} = \frac{2}{1} = 2$$

$$\tan 60° = \frac{O}{A} = \frac{\sqrt{3}}{1} = \sqrt{3} \approx 1.73 \qquad \cot 60° = \frac{A}{O} = \frac{1}{\sqrt{3}} = \frac{\sqrt{3}}{3} \approx 0.58$$

Unit Circle

With the **unit circle**, it is possible to find *trig values* for any size *angle*.

The unit circle has a *radius* of one unit.
It is set in an *x-y coordinate system*,
with its *center* at the *origin*.

What is a **radius**?

Radius is the same as **radio** and **radian,** and it comes from the same Latin word *radius*, which signifies **ray.** In geometry, a radius is a straight line from the center to the circumference, which is the boundary line of a circle.

In science, the concept of **radio** is of utmost importance in understanding the basic structure of all things — **the atom.** Two outstanding women scientists gave to the world an understanding of this minute, indivisible, homogeneous particle of which **all physical elements are composed of**. They are Marie Curie and Lise Meitner.

Marja Sklowdowska, later known as Marie Curie (1867-1934), has been one of the most remarkable physicists of all time. This Polish-born scientist **discovered the radioactive elements polonium and radium in 1902**, opening up the science of radioactivity, a word coined by her. She also described the behavior of uranium and thorium and was the first one to isolate the pure metal of radium. Nobody before her grasped the complex inner structure of the atom or the immense energy stored in it—the radioactivity energy.

Lise Meitner (1878-1968), an Austrian physicist, is a brilliant scientist who **discovered nuclear fission in the late 1930s.** She was the first person to verify that the uranium nucleus had split, rather than merely having small particles chipped off it, as it was thought before. Her observation led to the discovery of heavy nuclei fission, which was achieved by allowing a neutron to strike a nucleus of a fissile atom such as uranium-235 or plutonium-239. The nucleus of the fissile atom then splits apart to release two or three other neutrons. If the uranium-235 is pure, a chain reaction takes place when these neutrons in turn strike other nuclei. This happens very quickly, resulting in the tremendous release of energy seen in nuclear weapons. Meitner was invited to join the Manhattan Project to create the atomic bomb with a group of scientists, but she declined. She went to great lengths to distance herself from the negative possibilities her discoveries created.

Measurement of Angles in the Unit Circle

The unit circle is used as a powerful tool to measure all kinds of angles. It departs from a clear and simple structure—the x-y coordinate system with its center at the origin and a radius of one unit. The following is a whole circle, 360° with a total of 16 points, including the 4 points in the axes. These 16 pairs of coordinates automatically give the cosine and sine of the 16 angles.

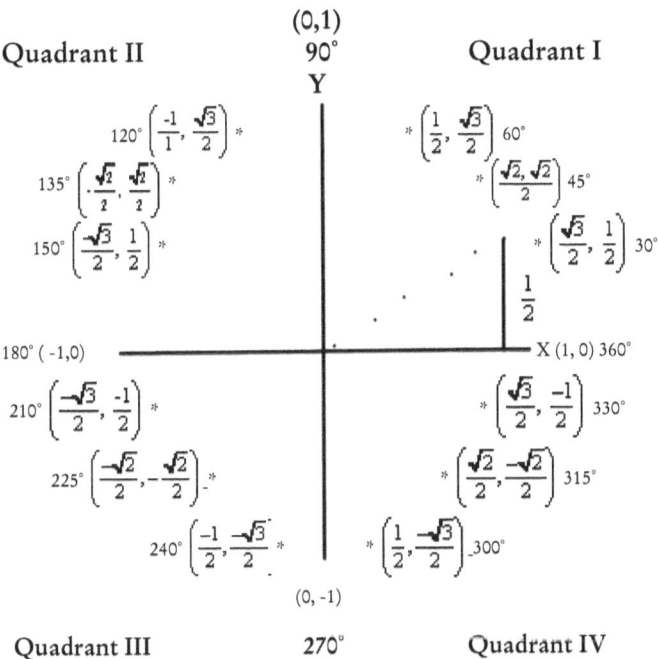

The starting point of the circle is 360°. If you go clockwise instead, you get an angle with a negative measure.

Notice that a triangle is enclosed within the circle. The cosine and sine are indicated in the first pair of the coordinates $\left(\frac{\sqrt{3}}{2}, \frac{1}{2}\right)$.

Hypotenuse:	1
Cosine of 30°	$\frac{\sqrt{3}}{2}$ = x-coordinate
Sine of 30°	$\frac{1}{2}$ = y-coordinate

Tangents are obtained by dividing an angle's sine by its cosine—in other words, the y-coordinate by its x-coordinate. And finally, the cosecant, secant, and cotangent of the 16 angles can easily be found because these trig functions are just the reciprocals of sine, cosine, and tangent.

Mnemonic—x and y (pair) are in alphabetical order as are cosine and sine.

There are two ways of measuring angles: degrees and radians. Radians *are determined on the length of the arc along the circumference of the unit circle cut off by the angle of the triangle. A radian is just a unit of angular measurement.* Radian comes from the Latin word *radian*, which means radius.

Circle Circumference

Degrees:	Radians:
360°	2 π radians
270°	1 ½ π radian $= \frac{3\pi}{2}$
180°	1 π radian

$90°$ ½ π radian $= \dfrac{\pi}{2}$

$60°$ ⅓ π radian $= \dfrac{\pi}{3}$

$45°$ ¼ π radian $= \dfrac{\pi}{4}$

$30°$ 1/6 π radian $= \dfrac{\pi}{6}$

Chapter 4

UNDERSTANDING THE POWER
OF CALCULUS

Calculus

C alculus is basically just **very advanced algebra and geometry**. It takes the ordinary rules of algebra and geometry and twitches them around so that they can be used in more complicated problems. **Differentiation** and **integration** constitute the core of the calculus curriculum.

Differentiation, is simply the process of finding the **derivative,** which is just a **rate,** like miles per hour or dollars per item. *Derivative* comes from the Latin word *derivare—de* meaning "from" and *rivus* meaning "river." It means taken from a source or origin. On the other hand, **integration,** is simply the process of bringing parts together into a whole.

Differential calculus is a system of mathematical analysis concerned with **finding the derivative** (rate of change) of a **variable function**.

Integral calculus is a system of mathematical analysis concerned with **finding the <u>limit</u> of a sum of terms** on an infinitesimal level.

In synthesis, to understand calculus is to understand that **the derivative** *is the* *inverse operation* **of the integral.**

Note: See Annex — Ten Important Graphs at the end of this book for further visualization.

Mathematics of Limits Formula

Three ways to find the *slope,* *area,* and *length* of a *curve* after zooming in "forever" to become perfectly straight at a microscopic level:

Differentiation	Integration	Arc length
——————	——————	——————
c . 3	. 3	. 3
——————	——————	——————
4	4	4
Slope	*Triangle Area*	*Hypotenuse* **Length**
$\dfrac{\text{rise}}{\text{run}} = \dfrac{3}{4}$	$\dfrac{1}{2} \text{ base x height} = 6$	$(a^2 + b^2 = c^2) = 5$

Limits and Continuity

Limits and continuity are fundamental for both differential and integral calculus, but with experience to use the shortcuts for calculating derivatives and integrals, there is no need to use the limit methods anymore.

The **limit** of a function for some x-value, **a,** is just when the **function height** (y-coordinate) gets closer and closer to **a** at the same time x gets closer and closer to **a** from the left and the right. The **limit** equals the height of the **gap**. An infinitesimal **gap** in a function is the only place a function can have a **limit** and where it is no longer a **continuous function**.

The **continuity** of a function for some x-value, **a,** is just that there is no **gap** there. **Gap** simply means a **discontinuity** (separation, rapture). In other words, a **continuous function is a function with no gaps** — a function that can be drawn without taking the pencil off the paper.

Definitions

Definition of Limit

$$\lim_{x \to a} f(x) *$$ exists if and only if: $$\lim_{x \to a-} f(x) = \lim_{x \to a+} f(x)$$

Definition of Continuity

$$\lim_{x \to a} f(x) *$$ exists if and only if: $$f(a) = \lim_{x \to a} f(x)$$

*   Let f be a function and **a** be a real number. A function $f(x)$ is *continuous* at a point x = a, if $f(a)$ is defined and the above two conditions are satisfied.

L'Hopital's Rule

Let f and g be differentiable functions. If the limit of $\frac{f(x)}{g(x)}$ as x approaches c produces $\frac{0}{0}$ or $\frac{\pm \infty}{\pm \infty}$ when the value of **c** is substituted into **x**, then

$$\lim_{x \to 0} \frac{f(x)}{g(x)} = \lim_{x \to c} \frac{f'(x)}{g'(x)}$$

*   c can be a number or $\pm \infty$

Example:

$$\lim_{x \to 3} \frac{x^2 - 9}{x - 3} = \lim_{x \to 3} \frac{2x^*}{1}$$

The new limit is:

$$\lim_{x \to 3} \frac{2x}{1} = \frac{2 \cdot 3}{1} = 6$$

*   Derive numerator and denominator separately: The derivative of $x^2 - 9$ is **2x**

The derivative of $x - 3$ is **1**

Note that -9 and -3 are *constants* so they are disregarded.

Solving Limit Problems with Algebra

Evaluate $\lim\limits_{x \to 5} \dfrac{x^2 - 25}{x - 5}$

1. **Factor:** $\lim\limits_{x \to 5} \dfrac{x^2 - 25}{x - 5} = \lim\limits_{x \to 5} \dfrac{(x - 5)(x + 5)}{x - 5}$

2. **Cancel x – 5:** $= \lim\limits_{x \to 5} (x + 5)$
   (from numerator/denominator)

3. **Substitute:** $= (5 + 5)$

                 $= 10$

Answer: $\lim\limits_{x \to 5} \dfrac{x^2 - 25}{x - 5} = 10$

Evaluate: lim $\quad \lim\limits_{x \to 4} \dfrac{\sqrt{x} - 2^{*}}{x - 4}$

* For square root functions, conjugate to get rid of square roots.

1. **Conjugate* and multiply:**

$$= \lim\limits_{x \to 4} \frac{(\sqrt{x} - 2)}{(x - 4)} \cdot \frac{(\sqrt{x} + 2)}{(\sqrt{x} + 2)}$$

$$= \lim\limits_{x \to 4} \frac{(\sqrt{x})^2 - 2^2}{(x - 4)(\sqrt{x} + 2)}$$

$$= \lim\limits_{x \to 4} \frac{(x - 4)}{(x - 4)(\sqrt{x} + 2)}$$

* Conjugation is to switch (-) for (+) or vice versa and place it in the numerator and denominator. Conjugates' product *always* = first term squared *minus* second term squared.

2. **Cancel (x - 4):** $\quad = \lim\limits_{x \to 4} \dfrac{1}{\sqrt{x} + 2}$

3. **Substitute:** $\quad = \dfrac{1}{\sqrt{4} + 2}$

$$= \dfrac{1}{4}$$

Answer: $\quad \lim\limits_{x \to 4} \dfrac{(\sqrt{x} - 2)}{x - 4} = \dfrac{1}{4}$

Note: The function $\dfrac{(\sqrt{x} - 2)}{x - 4}$ has a *hole* at $(4, \frac{1}{4})$

Evaluate:

$$\lim_{x \to 0} \frac{\dfrac{1}{x+4} - \dfrac{1}{4}^{*}}{x}$$

*   For complex fractions, multiply numerator and denominator by LCD, which is $4(x+4)$.

$$\lim_{x \to 0} \frac{\dfrac{1}{x+4} - \dfrac{1}{4}}{x} \cdot \frac{4(x+4)}{4(x+4)}$$

$$\lim_{x \to 0} \frac{4 - (x+4)}{4x(x+4)}$$

1. **Simplify:**

(cancel $+4 - 4$) $= \lim_{x \to 0} \dfrac{-x}{4x(x+4)}$

(cancel x) $= \lim_{x \to 0} \dfrac{-1}{4(x+4)}$

2. **Substitute:**

$$= \frac{-1}{4(0+4)}$$

Answer: This is the limit $= \dfrac{-1}{16}$

Limits at Infinity

Solve

$$\lim_{x \to \infty} \sqrt{x^2 + x} - x$$

*   Convert this function into a fraction by placing it over number 1 to conjugate.

1. **Conjugate/multiply:** $= \lim_{x \to \infty} \sqrt{x^2 + x} - x$

$$\lim_{x \to \infty} \frac{\left(\sqrt{x^2 + x} - x\right)}{1} \cdot \frac{\left(\sqrt{x^2 + x} + x\right)}{\left(\sqrt{x^2 + x} + x\right)}$$

$$\lim_{x \to \infty} \frac{\left(\sqrt{x^2 + x} - x^2\right)}{\left(\sqrt{x^2 + x} + x\right)}$$

$$\lim_{x \to \infty} \frac{x}{\left(\sqrt{1 + \frac{1}{x}} + 1\right)}$$

2. Factor x out of denominator:

$$= \lim_{x \to \infty} \frac{1}{1 + \frac{1}{x} + 1}$$

$$= \frac{1}{\sqrt{1 + \frac{1}{\infty}} + 1}$$

3. Substitute:

$$= \frac{1}{\sqrt{1 + 0} + 1}$$

$$= \frac{1}{1 + 1}$$

$$= \frac{1}{2}$$

Answer: $= \quad \lim_{x \to \infty} \left(\sqrt{x^2 + x} - x\right) = 1/2$

Limits to Remember

$$\lim_{x \to 0+} \frac{1}{x} = \infty$$

$$\lim_{x \to 0-} \frac{1}{x} = -\infty$$

$$\lim_{x \to \infty} \frac{1}{x} = 0 \qquad \lim_{x \to 0} \frac{\sin x}{x} = 1$$

$$\lim_{x \to -\infty} \frac{1}{x} = 0 \qquad \lim_{x \to a} c = c$$

$$\lim_{x \to 0} \frac{\cos x - 1}{x} = 0 \qquad \lim_{x \to \infty} \left(1 + \frac{1}{x}\right)^x = e$$

Note: The letter c means constant, which is any number. Number $e \approx 2.7182 \ldots$ It is the logarithm base and the most important number in calculus.

Series

Divergent/infinite series. Series of doubling numbers is divergent because if you continue the addition indefinitely, the sum will grow without limit. And if you could add up all the infinitely many **numbers,** **the sum of all these "infinitely many _numbers_" is** *infinity.* Divergent in most cases means that the series add up to infinity—that is,

$$1, 2, 4, 8, 16, 32, 64, \ldots$$

Convergent series. Series of convergent numbers are different from divergent because even though you keep adding more numbers and the sum keeps growing and the numbers and the sum grow forever, **the sum of all the "infinitely many _terms_" is a** *finite number.*

Differentiation

Differential calculus is the mathematics of **infinitesimal changes.**

Differentiation is the process of finding the **derivative**:

| | 3 feet | = rise (y coordinate) |

s   1 foot   = run (x coordinate)

Steepness/Slope/Derivative/rate/ratio $= \dfrac{\text{rise}}{\text{run}} = \dfrac{3}{1} = 3$

English   =   steepness
Algebra   =   slope
Calculus   =   derivative $= \dfrac{dy}{dx} = y'$

Note: Don't be confused with all the different symbols for the **derivative**.

They all mean <u>exactly the same</u>; $\dfrac{dy}{dx}$ and **y'** are the most commonly used.

$$\frac{dy}{dx} = y' \text{ or } \frac{df}{dx} \text{ or } \frac{df(x)}{dx} \text{ or } \frac{dy}{dx} \text{ or } f'(x) \underline{\text{ or }} Dxf \text{ or } Df \text{ or } Dxy \text{ or } \underline{Dxf(x)}$$

The symbol for the derivative or slope $\dfrac{dy}{dx}$ means a little bit of y to a little bit of x, as the slope shrinks down to an infinitesimal size. In other words as Δx approaches zero, $\dfrac{dy}{dx} = \dfrac{\Delta y}{\Delta x}$

Slope of Linear Functions
(Straight Lines with Unchanging Slopes)

Algebra deals with linear functions with *unchanging slopes*. A regular algebra problem gives the **average rate**.

Slope of Linear Functions Formula

$$\text{Slope} = \frac{y2 - y1}{x2 - x1}$$

Note: The linear functions with unchanging slopes as $y = 2x + 3$ is said to be in **slope-intercept form** because the **slope** and **y-intercept** is implicit in the equation:

$$y = mx + b$$

Equation: y = 2 x + 3 (where **m** is the **slope (2)** and **b (3)** is y-intercept when **x** is zero)

Evaluate: y = 2x + 3 *

1) Take any two points on the line, say (1,5) and (6,15).
2) Divide to get the slope.

$$\text{Slope:} \quad \frac{\text{rise}}{\text{run}} = \frac{15 - 5}{6 - 1}$$

$$= \frac{10}{5}$$

$$= \mathbf{2}$$

Answer: Unchanging slope is **2**.

* This function is in a *slope-intercept form*, which means that the **slope (2)** and the **y-intercept (3)** appear in the function. After graphing the function, different points appear as well as the **unchanging slope of 2.**

Points on the Line, y = 2x + 3, and the Slope of Those Points

x = run (horizontal)	1	2	3	4	5	6	7	8	etc.
y = rise (vertical)	5	7	9	11	13	15	17	19	etc.
Slope	2	2	2	2	2	2	2	2	etc.

Note: In addition to the *slope-intercept form*, the **point slope form** becomes handy to convert an explicit function into an implicit function (equation that defines **y** explicitly or implicitly as a function of **x**). For example, take the point **(6,15)** and plug it into the *point slope form*.

General *point slope form:* y - y1 = m (x - x1)

Implicit equation: y - 15 = 2 (x - 6) is converted into y = 2x + 3
                            y - 15 = 2x - 12 - 15
Explicit equation: y = 2x + 3

Derivatives of a Curve
(Parabolas, Curves with Changing Derivatives)

Calculus deals with parabolas, which are curves with changing derivatives. A regular calculus problem gives the **instantaneous rate**.

Difference Quotient Formula

$$f'(x) = \lim_{h \to 0} \frac{f(x+h) - f(x)}{h}$$

Difference quotient is just a fancy calculus term for the general slope

$$\frac{\text{rise}}{\text{run}} \quad \text{or} \quad \frac{y2 - y1}{x2 - x1}.$$

A **quotient** is a *fraction,* and both y2 – y1 and x2 – x1 are **differences.**

Evaluate: $y = \dfrac{1x^2}{4}$

1. Take the power and put it in front of the coefficient, and reduce the power by one to find the derivative.

$$\frac{1x^2}{4} \quad = \quad 2\frac{1}{4}x \quad \text{(the power 1 is not shown)}$$

$$= \quad \frac{2}{1} \cdot \frac{1}{4}x \ = \ \frac{2}{4}x$$

2. Reduce = 1x

Answer: The changing derivative is $\dfrac{dy}{dx} \ = \ \dfrac{1}{2}x \ $ or $\ y' \ = \ \dfrac{1}{2}x$

Points on the Parabola, $y = \dfrac{1}{4}x^2$, and its derivatives

x = run (horizontal)	1	2	3	4	5	6	etc.
y = rise (vertical)	0.25	1	2.25	4	6.25	9	etc.
Derivatives	0.5	1	1.5	2	2.5	3	etc.

To compute a derivative, two points are needed to plug into the formula. For a straight line, it is easy. But with parabolas, it is different, as is the case when a slope of a parabola is needed below at the point (2,4). There is a need of a **tangent line** intersecting the curve at **one point (2,4)**, and **secant lines** intersecting the parabola (curve) at **two points**, so the slopes of these secant lines get closer and closer to the slope of the tangent line because **the slope of the tangent line is precisely the derivative.**

Each secant line has a slope given by the formula $\dfrac{y2 - y1}{x2 - x1}$, which is the *average rate* over the interval from x1 to x2. The *instantaneous rate* is reached at the point (x1, y1) when the **limit** is taken and the slope of the **tangent line** is obtained.

Evaluate: $y = x^2$ with a tangent line at (2,4)

1. Bring the power of 2 in front and reduce the power by 1, which leaves a power of 1 that can be dropped because a power of 1 does nothing.

$$y = x^2$$
$$y' = 2^x$$

Eight Differentiation Rules

Constant Rule:		**Constant Multiple Rule:**	
If	$f(x) = c$ *	If $f(x)^3 = 4x\,2$	
Then	$f'(x) = 0$	Then $f'(x)^2 = 12\,x$	

Power Rule:

$$\text{If} \quad f(x) = x^{2}{}_{**}$$
$$\text{Then} \quad f'(x) = 2x$$

Product Rules:

If $\quad f(x) = x^{3} \bullet \sin x$

Then $f'(x) = 3x^{2} \sin x + x^{3} \cos x$

Note: Derivative of tangent

Quotient Rule:

If $\quad f(x) = \dfrac{\sin x}{x^{4}}$

Then $f'(x) = \dfrac{x \cos x - 4 \sin x}{5}$

If $\quad \tan x = \dfrac{\sin x}{\cos x}$

Then $(\tan x)' = \sec^{2}$

Chain Rule:

If

Then

$$f(x) = \sqrt{4x^{3}} - 5$$
$$f'(x) = 6x^{2}(4x^{3} - 5)^{-\frac{1}{2}}{}_{*}$$

Note: The *composite square function* has two functions (there is one function (4x − 5) inside another function (the square root function $\sqrt{}$).

The first function is **if** $f(x) = \sqrt{x}$ (same as x) **then** $f'(x) = \frac{1}{2}^{-\frac{1}{2}} x$

Then the second function is **if** $f(x) = 4x^{3} - 5$ **then** $f'(x) = 12x^{2}$

Simplify: $f'(x) = \frac{1}{2}(4x^{3} - 5)^{-\frac{1}{2}}(12x^{2}) = f'(x) = 6x^{2}(4x^{3} - 5)^{-\frac{1}{2}}$

Note: $f'(x) = 6x(4x - 5)$, which has a *negative fraction power* can be written with *positive fraction power* or with *no fraction power.*

Negative fraction power: *Positive fraction power:*   *No fraction power:*

$$f'(x) = 6x^2(4x^3 - 5^{-\frac{1}{2}}) \quad \text{or} \quad f'(x) = \frac{6x^2}{(4x^3 - 5^{\frac{1}{2}})} \quad \text{or} \quad f'(x) = \frac{6x^2}{\sqrt{4x^3 - 5}}$$

Sum Rule: Difference Rule:

If $f(x) = x^6 + x^3 + x^2 + x + 10$ Note: Just change the sign (+) for (-).

Remains the same as Sum Rule

Then $f'(x) = 6x^5 + 3x^2 + 2x + 1$

Note: Drop constant 10

*   c = **constant,** any ordinary number,   so if $f(x) = c$ then $f'(x) = 0$

k = **constant,** any ordinary number,   so if $f(x) = k$ then $f'(x) = 0$

so if $f(x) = 5$ then $f'(x) = 0$

so if $f(x) = 2k^3$ then $f'(x) = 0$

π = **constant,** $\pi \approx 3.14$, so if $f(x) = \pi$ then $f'(x) = 0$

**   Power rule works for any power: positive, negative, fraction. For example,

If $f(x) = x^2$ then $f'(x) = -2x$ If $f(x) = x^{\frac{2}{3}}$ then $f'(x) = 2/3x$

If $f(x) = x$ then $f'(x) = 1$ (note variable x has a power of 1)

If $f(x) = 5x$ then $f'(x) = 5$ (note 5 times (x = 1) equals 5)

If $f(x) = \pi x$ then $f'(x) = \pi$ (note π is a number/constant)

If $f(x) = \dfrac{5x^{\frac{1}{3}}}{4}$ then $f'(x)$ $= \dfrac{1(5)x^{-\frac{2}{3}}}{3 \cdot 4} = \dfrac{5^{-\frac{2}{3}}}{12}$

If $f(x) = 2x - 3x^{\frac{2}{3}} + 4$ then $f'(x) = 2 - 2(3)x^{-\frac{1}{3}} = 2 - 6x^{-\frac{1}{3}}$

If $f(x) = 5x + 2k^3$ then $f'(x) = 5$

Note: Differentiation with $e \approx 2.71$, $\pi \approx 3.14$

The derivative of e^x is itself:

If $f(x) = e^x$ **Then** $f'(x) = e^x$

If $f(x) = \pi x$ **then** $f'(x) = \pi$

Four Differentiation Trig Functions

Function	Derivative	Function	Derivative
If sinx	Then cosx	If cosx	Then sinx
		If tanx	Then $\sec^2 x$
If cscx	Then -cscxcotx*	If secx	Then secxtanx **
If cotx	Then $-\csc^2 x$		

Note: There is a mnemonic for last four trig derivatives:

$$^* \quad \csc > -\csc < \cot$$

$$^{**} \quad \sec > \sec < \tan$$

Explicit and Implicit Differentiations

Explicit differentiation means that **y** is written explicitly as a function of **x**, and that the equation **is solved for y;** and therefore, y is by itself on one side of the equation. This function can be written either with **y** or **f(x)**. For example,

Explicit function: $y = x^2 + 5x$ or $f(x) = x^2 + 5x$

Implicit differentiation means that **y** is written implicitly as a function of **x**, and that the equation **cannot be solved for y;** and therefore, there are two functions on either side of the equation that need to be differentiated. These two functions look like:

1. Differentiate both sides:

$$y^5 + 3x^2 = \sin x - 4y^3 \rightarrow 5y^4 + 6x = \cos x - 12y^2$$

2. Collect **y-terms** on left side and **x-terms** on right side:

$$5y^4 + 12y^2 = \cos x - 6x$$

3. Factor out **y':**

$$y' (5y^4 + 12y)^2 = \cos x - 6x$$

4. Divide for final answer:

$$y' = \frac{\cos x - 6x}{5y^4 + 12y^2}$$

Note: Remember that **x** and **y** *always* go in alphabetical order: **x** first and **y** second as in:

Pair of coordinates (axes point): **(x,y)**

Logarithmic Differentiation

Logarithmic differentiation is as **simple** as regular differentiation. The only difference is that the derivative of

$$\ln f(x) = \frac{1}{f(x)} \cdot f'(x) \text{ or } \frac{f'(x)}{f(x)}$$

A logarithmic function is an exponential function with the x and y axes switched. Log base e (e ≈ 2.72) is written *ln* instead of **log**.

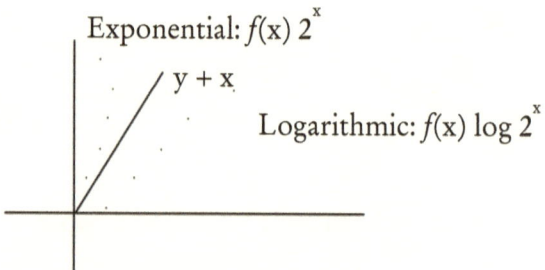

Exponential: $f(x)\, 2^x$

y + x

Logarithmic: $f(x)\, \log 2^x$

Regular Differentiation

Differentiate $f(x) = (x^3 - 5)(3x^4 + 10)(4x^2 - 1)(2x^5 - 5x^2 + 10)$
$f'(x) = 3x^2 + 12x^3 + 8x + 10x^4 - 10x$

Note: Remember that with regular differentiation, you always drop the **constants** as in the above operation: **-5, + 10, -1, + 10.**

Logarithmic Differentiation

Differentiate: $f(x) = (x^3 - 5)(3x^4 + 10)(4x^2 - 1)(2x^5 - 5x^2 + 10)$

1. Use the property for the log of the product.

$ln f(x) = ln(x^3 - 5)\, ln(3x^4 + 10)\, ln(4x^2 - 1)\, ln(2x^5 - 5x^2 + 10)$

2. Differentiate both sides according to the **chain rule,** the derivative of *ln*f(x) is:

$$ln\, f(x) = \frac{1}{f(x)} \cdot f'(x) \quad \text{or} \quad \frac{f'(x)}{f(x)}$$

$$f'(x) = 3x^2 + 12x^3 + 8x + 10x^4 - 10x$$

$$f(x) = (x^3 - 5)(3x^4 + 10)(4x^2 - 1)(2x^5 - 5x^2 + 10)$$

3. Finally, multiply both sides by $f(x)$ and you are finished.

$$f'(x) = \left[\frac{3x^2}{(x^3 - 5)} + \frac{12x^3}{(3x^4 + 10)} + \frac{8x}{(4x^2 - 1)} + \frac{10x^4 - 10x}{(2x^5 - 5x^2 + 10)} \right]$$

$$(x^3 - 5)(3x^4 + 10)(4x^2 - 1)(2x^5 - 5x^2 + 10)$$

Inverse Functions Differentiation

Inverse functions are symmetrical with respect to the line $y = x$. As with any pair of inverse function, if the point (4,10) is on one function, (10,4) is on its inverse.

The following general formula says that the **derivative** of a function, **f,** with respect to x is the reciprocal of the **derivative of its inverse (g)** with respect to **f.**

Inverse Functions Formula

$$f'(x) = \frac{1}{g'(f(x))}$$

Higher Order Derivatives

Finding second, third, fourth, or higher derivative is quite simple. The second derivative of a function is just the derivative of its first derivative. The third derivative is the derivative of its second derivative, and so on. For example,

$$
\begin{aligned}
f(x) &= x^4 - 5x^2 + 12x - 3 \\
f'(x) &= 4x^3 - 10x + 12 \\
f''(x) &= 12x^2 - 10 \\
f'''(x) &= 24x \\
f^{(4)}(x) &= 24 \\
f^{(5)}(x) &= 0 \\
f^{(6)}(x) &= 0 \\
\text{etc.} &= 0
\end{aligned}
$$

The higher derivatives of sine and cosine are cyclical:

$$
\begin{aligned}
f(x) &= \sin x \\
f'(x) &= \cos x \\
f''(x) &= -\sin x \\
f'''(x) &= -\cos x
\end{aligned}
$$

The cycle repeats indefinitely with every multiple of four.

Note: The first derivative of position is   >   **velocity**
(It tells how fast a function is changing with the slope going up and down.)

The second derivative of position is   >   **acceleration**
(It tells how fast the first derivative is changing, the rate of change of slope.)

Locating *Extrema*
f (x) Interpreting *Concavity* and *Points of Inflection*

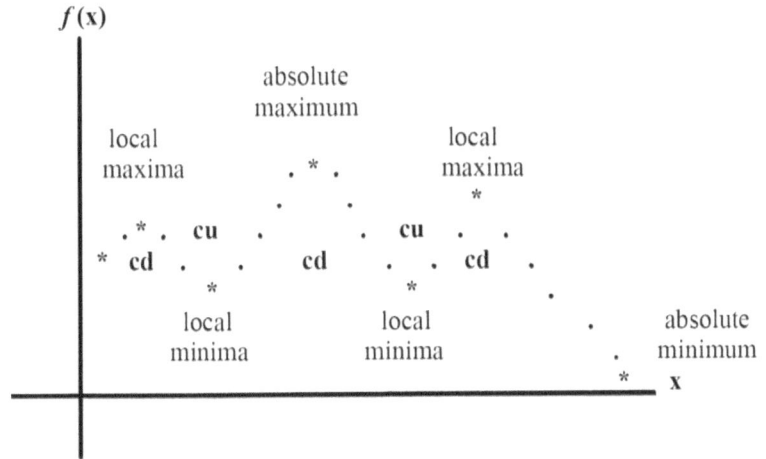

Locating *extrema*, simply means locating <u>all critical points</u> as depicted above. Critical points of the function are when its derivative is zero or undefined. There is **no** *tangent line* (**no** derivative) when there is a sharp turning point like a *cusp*. **Stationary points** are called *all local maxima points*.

Interpreting concavity simply means locating <u>*concave down*—cd</u> (down the curve) or <u>*concave up*—cu</u> (up the curve). **Inflection points** are called when conc*avity changes down to up.*

Note: When the curve is going *down* — function is decreasing and **its derivative (slope) is negative**

When the curve is going *up* — function is increasing and its derivative (slope) is positive

Finding Extrema

Find extrema (critical points) of $f(x) = 3x^5 - 20x^3$

1. Differentiate: $f(x) = 3x^5 - 20x^3$

   $f'(x) = 15x^4 - 60x^2$

2. Set the derivative ($f'(x)$), left side of the equation to **zero,** and solve for **x.**

$$15x^4 - 60x^2 = 0$$
$$15x^2 - (x - 4)^2 = 0$$
$$15x^2 - (x + 2)(x-2) = 0$$

$$15x^2 = 0 \quad \text{or}$$
$$x + 2 = 0 \quad \text{or}$$
$$x - 2 = 0 \quad \text{so}$$

Answer: x = 0, 2, or-2

Once you have the *list of the critical points*, you can double-check whether *the list of the critical points* actually occur at those x values. You can do this with either the **first derivative test** or the **second derivative test.**

First Derivative Test — Function Values

1. Draw a number line and put down the **critical points** you found above:

0, 2, or -2

The Critical Points of $f(x) = 3x^5 - 20x^3$

Critical numbers

3. Divide the above number line into **four regions**:

to the left of -2, from -2 to 0, from 0 to 2, and to the right of 2.

3. Now pick a value from each region, plug it into the first derivative, and note whether the result is positive or negative. Let's use **-3, -1, 1, and 3** to test the regions.

$$f'(x) = 15x^4 - 60x^2$$

$f'(-3) = 15(-3)^2 - 60(-3)^2$
$= 15 \cdot 81 - 60 \cdot 9$
$= 675$

$f'(-1)^4 = 15(-1)^2 - 60(-1)$
$= 15 - 60$
$= -45$

$f'(1) = 15(1)^4 - 60(1)^2$
$= 15 - 60$
$= -45$

$f'(3) = 15(3)^4 - 60(3)^2$
$= 15 \cdot 81 - 60.9$
$= 675$

Sign Graph and the Critical Numbers of:
$$f(x)\ 3x^5 - 20x^3$$

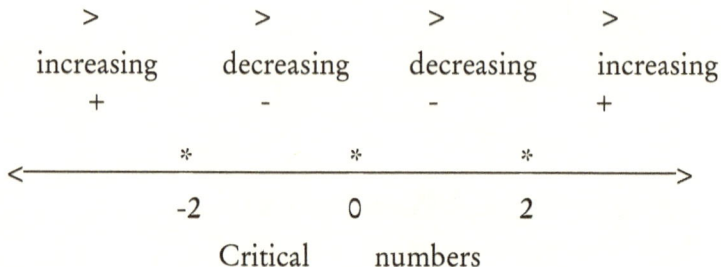

>   >   >   >

increasing   decreasing   decreasing   increasing

\+   -   -   \+

      *   *   *

$\longleftarrow$ ——————————————— $\longrightarrow$

     -2   0   2

Critical   numbers

Function Values (heights)

$$f(x)\ =\ 3x^5 - 20x^3$$

$$f(-2)\ =\ 3(-2)^5 - 20(-2)^{3*}$$

$$=\ 64$$

$$f(2)\ =\ 3(2)^5 - 20(2)^3$$

$$=\ -64$$

Thus, **the local max** is located at   (-2, 64), and

the local min is located at   (2, -64).

(x = run, y = rise (height)

*   3 and 20 are *coefficients* in the algebraic expressions; 3(-2)^5 and -20(-2)^5 are **not multiplied until the number raised to a certain power is resolved.**

Therefore, $3(2)^5$ is solved by first resolving:

$$(-2)^5$$
$$= (-2)(-2)(-2)(-2)(-2)$$
$$= -32$$

then $+ 3(-32) \qquad = -96$

Therefore, $-20(-2)$ is solved, by first resolving:

$$(-2)^3$$
$$= (-2)(-2)(-2)$$
$$= -8$$

then $-20(-8) \quad = +\dfrac{160}{64}$

Note: signs $+$ and $-$ $\quad = -$

$\qquad\qquad -$ and $-$ $\quad = +$

$\qquad\qquad +$ and $+$ $\quad = +$

Second–Derivative Test — Function Concavities

$f(x) \qquad = \qquad 3x^5 - 20x^3$

$f'(x) \qquad = \qquad 15x^4 - 60x^2 \qquad$ (first derivative)

$f''(x) \qquad = \qquad 60x^3 - 120x \qquad$ (2nd derivative)

$f''(-2) \qquad = \qquad 60(-2)^3 - 120(-2) \ = \ -240$

$f''(-2) \qquad = \qquad 60(0) - 120(0) \ = \ 0$

$f''(-2) \qquad = \qquad 60(2)^3 - 120(2) \ = \ 240$

Thus, $(x, y) = (-2, -240)$ tells that:

At -2, the second derivative of $\quad$ f (function /y) is negative $= -240$

$\qquad\qquad\qquad\qquad\qquad$ Concave down, where x $\quad = -2$,

$\qquad\qquad\qquad\qquad\qquad$ Local max at $\qquad\qquad\qquad$ -2.

At **2,** the second derivative of   f **(function/y) is positive**   = 240

                              Concave up, where   x = 2

                              Local min at   2

Mean Value for Continuous Function Theorem

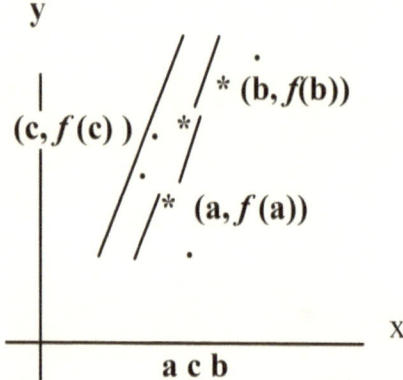

If f is a continuous function on the closed interval [a,b] and differentiable on the open interval (a,b), **then** there exists a number c in (a,b). This theorem means that the secant line connecting points $(a, f(a))$ and $(b, f(b))$ has a slope given by the slope formula:

$$\text{Slope} = \frac{y2 - y1}{x2 - x1}$$

$$= \frac{f(b) - f(a)}{b - a}$$

8 Most Common Differential Problems

1) Maximum   Area
2) Maximum   Volume
3) Radius   Rate
4) Related   Rate
5) Tangent   Line
6) Normal   Line
7) Position   Velocity and Acceleration
8) Marginal   Cost, Revenue, Profit

Maximum Area

Maximum Area of a Rectangle Corral

Solve the maximum area of a corral that is divided into 2 equal rectangles with 300 feet of fencing:

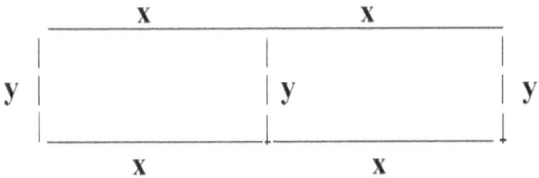

Note that the fencing is used for seven sections — sections of length marked with **x** and sections of width marked with **y**. (4 sections x and 3 sections y)

$$300 = x + x + x + x + y + y + y$$
$$300 = 4x + 3y$$

1) **Express the area to be maximized** as a function of two unknowns, x and y, and solve it for **y:**

$$
\begin{aligned}
A &= 1 \cdot w \\
A &= (2x)(y)* \\
4x + 3y &= 300 \\
3y &= 300 - 4x \\
y &= \frac{300 - 4x}{3} \\
y &= 100 - \frac{4x}{3} \\
{}^*A &= (2x)(y)\left(100 - \frac{4x}{3}\right) \\
A(x) &= (2x) \\
A(x) &= 200x - \frac{8x^2}{3}
\end{aligned}
$$

2) **Determine the function domain,** *sections length,* which in this case is:

$$300 \div 4 = 75$$

3) **Find the critical numbers** of A (x) in the open interval (0,75) by setting its derivative to zero and solve for **x:**

$$
\begin{aligned}
A(x) &= 200x - \frac{8x^2}{3} \\
A'(x) &= 200 - \frac{16x}{3} \quad \text{(power rule)} \\
200 - \frac{16x}{3} &= 0 \\
-\frac{16x}{3} &= -200 \\
x &= -200 \cdot \frac{(-3)}{16} \\
&= \frac{600}{16} \\
&= 37.5
\end{aligned}
$$

Note: Because A is defined for all x values, 37.5 is the only critical number. The maximum value in the interval is **3,750**, and thus, an x value of 37.5 feet maximizes the corral's area. The length is 2x or **75 feet.** The width is y, which equals 100 – 4 (37.5) or **50 feet.**

Answer: The maximum area of this corral is 3,750 square feet. The corral's dimensions are:

$$\text{Area} = \text{length} \cdot \text{width}$$

$$3{,}750 \text{ sq.feet} = 75 \text{ feet} \cdot 50 \text{ feet}$$

Maximum Volume of a Square Box

Solve the maximum volume of a box with the following dimensions:

$$(V) = \text{Volume} - \text{maximum}$$
$$(l) = \text{Length} - 30 \text{ inches}$$
$$(w) = \text{Width} - 30 \text{ inches}$$
$$(h) = \text{Height} - 2 \text{ inches}$$

Note that there are two variables **V** for volume and **h** for height because they are both changing.

1) **Express the volume to be maximize** as a function of one unknown (*the height of the box*) and **factor:**

$$V = l \cdot w \cdot h$$

$$V(h) = (30 - 2)(30 - 2) \cdot h^{*}$$

$$= \quad (900 - 120h + 4h)^2 \cdot h \text{ (factoring)}$$

$$= \quad 4h^3 - 120h^2 + 900h$$

*   **Remember FOIL:**

First	multiply:	30 x 30	=	900
Opposite	multiply:	30 x 2h	=	60h
Interior	multiply:	2 x 30	=	$\dfrac{60h}{120h}$
Last	multiply:	2h x 2h	=	**4h**

2) **Determine the function domain,** *maximum height***, which in this case is **15.**

$$30 \div 2 = 15$$

**   Maximum height is half the width of the box.

3) **Find the critical numbers** *of V(h) in the open interval* (0,15) by setting the **derivative to zero and solve:**

V (h)	=	$4h^3 - 120h^2 + 900h$	
V' (h)	=	$12h^2 - 240h + 900$	
0	=	$12h^2 - 240h + 900$	(power rule)
0	=	$h^2 - 20h + 75$	(dividing by 12)
0	=	(h − 15) (h − 5)	(factoring)
h	=	**15 or 5**	

Note: **15** is the *maximum height* and **5** is the *critical point,* so

Height of **5** inches produces the box with maximum volume.
Length and width = 30-2. **5** = 30 -10 = **20**

Answer: The maximum volume of this square box is **2,000 cubic inches.** Its dimensions are:

Volume = length. width. height

2,000 cubic inches = 20"   •   20"   •   5"

Radius *Rate*

Geometry Formula for the Sphere Volume:

$$V = \frac{4}{3}\pi r^3$$

Radius Increasing Rate in a Balloon

Solve the *radius increasing rate* in a balloon when its radius is 3 inches and it has been blown at a rate of 300 cubic inches per minute.

Note that *radius* is labeled with the variable **r,** and the *volume* is labeled with the variable **V** because both are *changing.* Notice also that $\frac{dV}{dt} = 300^*$

1) **Differentiate** with respect to time **t**:

$$\frac{dV}{dt} = \frac{4}{3}\pi \cdot 3r^2 \frac{dr}{dt}$$

$$= 4\pi r^2 \frac{dr}{dt}$$

2) **Substitute** <u>dV</u> for 300*:

$$300 = 4\pi \cdot 3^2 \frac{dr}{dt}$$

$$300 = 36\pi \frac{dr}{dt}$$

$$\frac{300}{36\pi} = \frac{dr}{dt}$$

$$\frac{dr}{dt} \approx 2.65 \text{ inches per minute}$$

Answer: The radius is increasing at a rate of about 2.65 *inches per minute* when the radius measures 3 inches.

Related *Rate*

Geometry Formulas

Right Prism   All triangles

$$V = A \bullet h \qquad\qquad A = \frac{1}{2} b \bullet h$$

(A is the area of the base)

Related Rate of a Vessel

Solve how fast the water level rises when the depth of a vessel is 1 foot 3 inches, and it is being filled with water at a rate of 5 cubic feet per minute. The vessel is 10 feet long and its cross section is an isosceles triangle with a base of 2 feet and a height of 2 feet 6 inches.

Note that height is labeled with the variable **h**, the area of the base is labeled with the variable **A**, volume is labeled with the variable **V**, and time is labeled with the variable **t** because all of them are changing. Also notice that the length of the vessel is 10 feet long, so the formula becomes:

$$V = \frac{1}{2}\, bh \cdot 10 \quad \text{or} \quad V = 5bh\,*$$

And to suppress the variable **b,** we end up just with two variables **V** and **h:**

$$\textbf{b and h} \text{ are proportional so}: \qquad \frac{b}{2} = \frac{h}{2.5}$$

$$\text{Cross multiplication:} \qquad 2.5b = 2h$$

$$b = \frac{2h}{2.5}$$

$$b = 0.8h$$

Finally, notice substitution formula $\quad V = 5bh\,^*$

(5 times 0.8 = 4, h times h = h^2) $\qquad V = 5 \bullet 0.8h \bullet h$

$$V = 4h^2$$

1) **Differentiate** with respect to time **t:**

$$\frac{dV}{dt} = 8h\frac{dh}{dt}$$

2) **Substitute** and plug **5** and **1.25** to solve for $\frac{dh}{dt}$:

$$\frac{dV}{dt} = 5 \text{ cubic feet per minute}$$

$$h = 1 \text{ foot 3 inches or } \textbf{1.25 feet}$$
$$(1 \text{ foot} = 12 \text{ inches})$$

$$\frac{dV}{dt} = 8h\frac{dh}{dt}$$

(8 times 1.25 = 10) $5 = 8 \times 1.25 \times \dfrac{dh}{dt}$

(5 = 10 is equal to $\frac{1}{2}$) $5 = 10 \quad \times \quad \dfrac{dh}{dt}$

$$\frac{dh}{dt} = \frac{1}{2}$$

Answer: The water level is rising at a **rate** of <u>1</u> foot per minute when the water is 1 foot 3 inches deep. 2

Another Related Rate Problem

Pythagorean Theorem

$$a^2 + b^2 = c^2$$

Related Rate of Cars

Solve how fast the distance changes between two cars at a point when the northbound car is 3/10 or 0.3 of a mile north of an intersection, and the westbound car is 4/10 or 0.4 of a mile east of the same intersection. The first car leaves the intersection traveling north at 50 mph, and the second car is driving west towards the same intersection at 40 mph.

Note that there are four variables: **t** for time, **y** for the car speeding at 50 mph, **x** for the car speeding at negative – 40 mph, and **s** for the unknown rate; so using the Pythagorean theorem, it becomes:

$$a^2 + b^2 = c^2$$

$$x^2 + y^2 = s^2$$

1) **Differentiate** with respect to **t**:

$$s^2 = x^2 + y^2$$

(power rule) $\quad 2s \dfrac{ds}{dt} = 2x \dfrac{dx}{dt} + 2y \dfrac{dy}{dt}$

2) **Substitute** and solve for $\dfrac{ds}{dt}$:

$$s^2 = x^2 + y^2$$
$$s^2 = 0.4^2 + 0.3^2$$
$$= 0.16 + 0.09$$
$$= 0.25$$
$$S = 0.5\,^*$$

*   Taking the square root of both sides $\sqrt{0.25} = 0.5$

$$2s \frac{ds}{dt} = 2x \frac{dx}{dt} + 2y \frac{dy}{dt}$$

$$2 \cdot 0.5 \frac{ds}{dt} = 2 \cdot 0.4 \cdot (-40) + 2 \cdot 0.3 \cdot 50) *$$

$$1 \frac{ds}{dt} = -32 + 30$$

$$\frac{ds}{dt} = -2$$

*   (2 x 0.4 = 0.8 x -40 =-32)
    (2 x 0.3 = 0.6 x 50 = 30)

Answer: -2 means that the distance s is decreasing. Thus, when the first car is 3 blocks north of the intersection and the second car is 4 blocks east of the intersection, the distance is decreasing at a rate of **2 mph.**

Tangent Line

Determine Two Points of Tangency

Solve the location of the two points of tangency through the point (**1,-1**) that are tangent to the parabola, using the equation:

$$y = x^2$$

Note: Because the equation of the parabola is y = x, we can take a general point on the parabola, (x,y), and substitute

x for y, so $(x,y)^2$ becomes $(x,x)^2$

1) **Differentiate:** $y = x^2$

$y' = 2x$ (power rule)

2) **Substitute** and solve for **x:**

Use the slope formula: $\dfrac{y2 - y1}{x2 - x1}$ $\dfrac{x^2 - (-1)}{x - 1} = 2x$

(fraction with algebra terminology)

$$x^2 - (-1) = 2x\,(x - 1)$$

$$x^2 + 1 = 2x^2 - 2x$$

$$0 = x^2 - 2x - 1$$

Use the quadratic formula:

$$x = \dfrac{-b \pm \sqrt{b^2 - 4ac}}{2a} \qquad x = \dfrac{2 \pm \sqrt{(-2) - 4(1)(-1)} \ *}{2(1)}$$

* Substract $(-2)\,(-4) = (-2) \times (-2) = 4 = 2 \pm \sqrt{4 + 4}$

Multiply $(-4)\,(+1) = (-4) \times (-1) = 4$ ____

$$= \dfrac{2 \pm \sqrt{8}}{2}$$

$$= \dfrac{2 \pm 2\sqrt{2}}{2}$$

$$= 1 \pm \sqrt{2}$$

Therefore, the **x coordinates** of the points of tangency are:

$$1 + \sqrt{2} \quad \text{or} \quad 1 - \sqrt{2}$$

3) **Finally, plug** each of the **x coordinates** into $y = x^2$ to obtain the **y coordinates.**

$$y = (1+\sqrt{2})^2 \qquad\qquad y = (1-\sqrt{2})^2$$
$$y = 1+2\sqrt{2}+2 \qquad\qquad y = 1-2\sqrt{2}+2$$
$$y = 3+2\sqrt{2} \qquad\qquad y = 3-2\sqrt{2}$$

Therefore, the **two points of tangency are:**

$$\left(1+\sqrt{2},\ 3+2\sqrt{2}\right) \qquad \text{and} \qquad \left(1-\sqrt{2},\ 3-2\sqrt{2}\right)$$

$= 1+1.4 = 2.4$, $3+1.4 = 4.4+1.4$ and $=1-1.4 = 0.4$, $3-1.4 = -1.6-1.4$

or about　　　**(2.4,5.8)**[*]　　　　　　and　　　　　　**(-0.4, 0.2)**

[*]　Remember: $\sqrt{0} = 0$, $\sqrt{1} = 1$ $\sqrt{2} = 1.41$, $2\sqrt{2} = -1.41$ and, $2\sqrt{x^{-1}} = x^{\frac{1}{2}}$, $x^0 = 1$

(Zero raised to the zero power is **undefined.**)

Another Tangent Line Problem

Determine Tangent Line

Note: A line tangent to a function is a good approximation of the function near the **point of tangency.** The curve and the tangent line are virtually indistinguishable; the further they zoom in on them, the straighter they look.

Solve the equation of the line tangent of $f(x) = \sqrt{x}$ at $(9,3)$. The **slope** is given by the **derivative** of f at **9.**

1) **Differentiate:**

$$f(x) = \sqrt{x}$$
$$= x^{\frac{1}{2}}$$
$$f'(x) = \frac{1}{2}x^{\frac{1}{2}} \qquad \text{(power rule)}$$

2) Substitute:

$$f'(x) = \frac{1}{2\sqrt{x}}$$

$$f'(9) = \frac{1}{2\sqrt{9}}^*$$

$$= \frac{1}{6}$$

* Square root of $\sqrt{9}$ is 3 (3 x 3 = 9)

So **2 times 3 = 6**

The slope is $\frac{1}{6}$

3) Plug the slope $\frac{1}{6}$, the point (9,3), and 10 (an approximation to 9)

Use the point slope form:

$$y - y1 = m (x - x1)$$

$$y - 3 = \frac{1}{6} (x - 9)$$

$$y = 3 + \frac{1}{6} (x - 9)$$

$$y = 3 + \frac{1}{6} (10 - 9)$$

$$y = 3 + \frac{1}{6}$$

$$y = 3 \frac{1}{6}$$

Answer: The line tangent is $3 \frac{1}{6}$ —an approximation of the square root of 10.

Note: The **slope (ratio of the rise to the run)** of this function is given by the **derivative 9,** and it goes through the point (x,y) = **(9,3).**

This equation, $f(x) = \sqrt{x}$ at **(9, 3)** can be written now in the *slope-intercept form:*

y = mx + b form (where **m** is the *slope* and **b** is the **y-intercept),** thus:

Equation: $f(x) = \sqrt{x}$ at **(9, 3)**

Point slope form: $y = 3 + \frac{1}{6}(x - 9)$

Slope-intercept form: $y = \frac{1}{6} x + 3$

Approximation line and its *points*

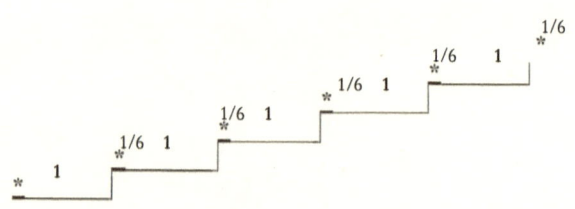

4/6	5/6		1/6	2/6	3/6
7,2	8,2	9,3	10,3	11,3	12,3

Normal *Line*

Determine points of perpendicularity for all *normal lines* to the parabola. *Normal lines* are the lines perpendicular to a curve at a given **tangent point**.

Note that the **slopes** of perpendicular lines, *normal lines*, are **opposite reciprocals**, which are the negative counterparts of a number/term.

Solve the location of points of $y = \dfrac{1}{16} x^2$ passing through the point **(3,15)**.

1) **Differentiate:**

$$y = \frac{1}{16} x^2$$

$$y' = \frac{1}{8} x$$

2) **Substitute** and solve for **x** by using the <u>opposite reciprocal</u>:

	Reciprocal	Opposite Reciprocal
$\dfrac{1}{8} x$ =	$\dfrac{8}{1} x$ =	$-\dfrac{8}{1} x$

Use the slope formula $\dfrac{y2 - y1}{x2 - x1}$:

Note: The opposite reciprocal of $\frac{1}{8}x$ or $\frac{x}{8}$ *is* $-\frac{8}{x}$

$$\frac{\frac{1}{16}x^2 - 15}{x - 3} = -\frac{8}{x}$$

Cross multiply and distribute $\qquad \frac{1}{16}x^3 - 15x = -8x + 24$

Bring all terms to left side, x by 16 $\qquad x - 112x^3 - 384 = 0$

(15x - 8x =   -7x (16)   = -112x

　　　and   -24 (16)   = -384)

Therefore, y = x^3 - 112x - 384

Note: There is no automatic way to get exact solutions to cubic, third-degree equations as the way the quadratic formula gives the solutions to quadratic, second-degree equations, but this particular case is different.

3) **Plug** each of the three **x** coordinates **-8, -4,** and **12** into **y =** $\frac{1}{16}x^2$ to get the **y** coordinates.

$$y = \frac{1}{16}(-8)^{2*}$$

[*$(-8)^2 = (-8)(-8) = 64$; $64 \div 16 = 4$] = 4 **

** $4 = \frac{4}{1}$; reciprocal is $\frac{1}{4}$; opposite reciprocal is $\frac{-4}{1}$ or (-4) $y = \frac{1}{16}(-4)^2$

$$= 1 ***$$

[$(-4)^2 = (-4)(-4) = 16$; $16 \div 16 = 1$]

*** $1 = \frac{1}{1}$; reciprocal is $\frac{1}{1}$; opposite reciprocal is $-\frac{1}{1}$ or -1

Note: Two previous points of normalcy are (x, y)

$$= (-8, 4)$$
$$= (-4, 1)$$

Third point is the sum of -8 + -4 for x = (12,) $y = \dfrac{1}{16} (12)^2$

$$= 9 \; ^{****}$$

**** $[(12)^2 = 12 \times 12 = 144; 144 \div 16 = 9]$

Answer: The three normal points on the parabola are (-8,4), (-4,1), and (12,9).

Position, Velocity, and Acceleration

Determine position, velocity, and acceleration.

Note: A function of time gives *position*, the first derivative gives its *velocity*, and the second derivative gives its *acceleration*.

Solve: $H(t) = t^3 - 6t^2 + 5t + 30$, where H = height in inches, t = time in seconds.

Function *position—height* (Ht):
Velocity (Vt) is derivative of (Ht):
Acceleration (At) is derivative of (Vt):

1) **Differentiate:** 3 2
   Use Power Rule

$$H(t) = t^3 - 6t^2 + 5t + 30$$
$$H'(t) = 3t^2 - 12t + 5$$
$$H''(t) = 6t - 12$$

2) Substitute to find maximum and minimum heights:

Use Quadratic Formula

$$x = \frac{-b \pm \sqrt{b^2 - 4ac}}{2a}$$

$$0 = 3t^2 - 12t + 5$$

$$t = \frac{-(-12) \pm \sqrt{(-12)^2 - 4(3)(5)}}{2 \cdot 3}$$

$$= \frac{12 \pm \sqrt{84}}{6} \quad *$$

* $\sqrt{84} = \sqrt{84}^{\frac{1}{2}} = 2\sqrt{84} = 2\sqrt{21}$

$$= \frac{12 \pm 2\sqrt{21}}{6} \quad \#$$

(½ (84) = 42; ½ (42) = 21)

$$= \frac{6 \pm \sqrt{21}}{3} \quad ***$$

** $\dfrac{12}{6} = \dfrac{6}{3}$

*** $\dfrac{6}{3} = (3 \times 2 = 6) = \dfrac{2}{1} = 2$ $\quad (3 \times 1 = 3)$

$$= \frac{2 \pm \sqrt{21}}{3}$$

$\sqrt{21} \approx (4.58 \times 4.58 \approx 20.97) \approx 4.58$

$\dfrac{\sqrt{21}}{3} \approx 4.58 \div 3 \approx 1.53$

**** 1.53 - 2.00 ≈ 0.47

$$= \sim 0.47 \text{ or } \sim 3.53 \quad ****$$

1.53 + 2.00 ≈ 3.53

Answer: This function gets as **high** as ≈ **31.1 inches** at ≈ **0.47 seconds** and as low as ≈ **16.9 inches at ≈ 3.53 seconds.** At **t = 0** it is **30** *inches high*, at **t = 4** it is 14 *inches high*.

16.9 inches at ≈ 3.53 seconds.

H(t) = *height in inches*

```
30 | *.
27 |      .
24 |        .
21 |
18 |          * (4,18)   =   (x,y)
15 |        .
12 |        .
 9 |        .
 6 |        .
_3 |_____  t = time in seconds
   | 1  2  3  4
```

Note: Fractions with square roots are *approximations*, and they are very simple. For example,

$\sqrt{3} \approx 1.73$ (1.73 x 1.73 ≈ 2.9929)   ≈ **3**

$\dfrac{\sqrt{3}}{2} \approx$ 1.73 divided by 2 ≈ 0.865   ≈ **0.87**

$\dfrac{\sqrt{3}}{3} \approx$ 1.73 divided by 3 ≈ 0.57666 . . .   ≈ **0.58**

$\dfrac{2\sqrt{3}}{3} =$ 1.73 divided by 3 = 0.58 + 0.58 = 1.16   = **1.15**

Velocity vs Speed

| Upward motion | = *positive velocity* | vs | **always** *positive speed* |
| Down/left motion | = *negative velocity* | | (marked by speedometer) |

Displacement vs Distance

| Down/left | = negative displacement | vs | **always** *positive distance* |
| (total: final position minus initial position) | | | (total: distance traveled) |

Therefore, in the previous **function**, we find six things:

Three things in *displacement and velocity:*

> *Negative total displacement*
> *Average velocity*
> *Maximum and minimum velocity*

Three things in *distance and speed:*

> *Total distance traveled*
> *Average speed*
> *Maximum and minimum speed*

The function has a *negative total displacement* because the position starts at a height of 30 inches and ends at a height of 18 inches.

Total displacement = 18 − 30 = **-12**

It is *negative* because the net movement is *downward.*

The function has an *average velocity* of 3 inches per second given by *total displacement* divided by *elapsed time.*

$$\text{Average velocity} = \frac{-12}{4}$$
$$= -3$$

It is *going down on an average* of 3 inches per second.

The function has a *maximum velocity of 5 inches per second* twice (one at t = 0 and another at t = 4) **and** *minimum velocity of -7 inches per second* at t = **2** seconds during the interval from 0 to 4 seconds (*elapsed time is 4 seconds*) at points (0,5), (2,-7), and (4, 5).

First derivative (velocity)Second derivative (acceleration)

$$H' = 3t^2 - 12t + 5 \qquad\qquad H'' = 6t - 12$$

t(0) = 5	6t - 12 = 0
t(2) = -7	6t = 12
t(4) = 5	t = 2

H'(t) = high in inches (velocity)

t(0) = **5** (0,5)   t(4) = **5** (4,5)

t = **time in seconds** (0 to 4 sec.)

t(2)= –7 (2, –7)

It has a *total distance* traveled **of 16.4 inches.**

Total distance = ~1.1 + ~ 14.2 + ~ 1.1 ≈ 16.4 inches

From the initial position of 30 inches, it goes about 31.1 inches, which is a distance of about **1.1 inches.** Then it goes from 31.1 inches to about 16.9 inches, which is a distance of about **14.2 inches.** Finally, it goes up again from approximately 16.9 inches to its final height of 18 inches, which is another **1.1 inches.**

The function has an *average distance* of **4.1 inches per second** given by the total distance traveled divided by the elapsed time of **4 seconds** (*interval time from 0 to 4 seconds*).

$$\text{Average speed} \approx \frac{16.4}{4} \approx 4.1 \text{ inches per second}$$

It has a *maximum speed* of *7 inches per second* and its *minimum speed* is *0.*

Maximum velocity

Acceleration and Deceleration

Acceleration is just a change in speed per unit of time so-called *second squared.*

The function has periods of acceleration and deceleration:

Acceleration (positive acceleration) from t = 2 to t = 4 **Velocity increasing**	Deceleration (Negative acceleration) from t = 0 to t = 2 **Velocity decreasing**

It has a maximum acceleration of *12 inches* at the far right when *t = 4* and **minimum acceleration of -12 inches** at the far left when *t = 0.*

1) Differentiate:

$$A(t) = 6t - 12$$
$$A(t') = 6$$
$$0 = 6$$

2) Absolute extrema occurs at the interval's endpoints, 0 and 4

$$A(0) = 6 \cdot 0 - 12$$

$$= -12 \quad \frac{\text{inches per second}}{\text{second}}$$

$$A(4) = 6 \cdot 4 - 12$$

$$= 12 \quad \frac{\text{inches per second}}{\text{second}}$$

Business and Economic Problems

(Marginal Cost, Marginal Revenue, and Marginal Profit)

Determine cost, revenue, and maximum profit margins of a manufactured gadget.

Note: Margins in economics *work exactly the same way as the linear approximation.* For instance, a manufacturer determines that the demand function for her gadgets is

$$p = \frac{1000}{\sqrt{x}}$$

where p = price and x = demand of gadgets at a given price.

Solve the *marginal cost* function: $C(x) = 10x + 100\sqrt{x} + 10,000$, and evaluate it at x = 100. C stands for cost.

Differentiate:

$$C(x) \quad = \quad 10x + 100\sqrt{x} + 10,000*$$

$$C'(x) \quad = \quad 10 + \frac{50}{\sqrt{x}}$$

$$C'(100) \quad = \quad 10 + \frac{50}{\sqrt{100}}$$

(50 divided by 10 = 5 +10 = 15) $= \quad 10 + \frac{50}{10}$

$$= \quad 15$$

* **Remember:** The power rule disregards the *constants*, in this case 10,000, and reduces 100 by half = 50.

Answer: The approximate *marginal cost* at x = 100 is **$15** for the first 10 gadgets.

Solve the *marginal revenue*, where R stands for revenue, p stands for price, and R(x) equals the number of items sold (x times p).

$$R(x) = x \cdot p$$

$$= x \cdot \frac{1,000}{\sqrt{x}}$$

$$= \frac{1,000x}{\sqrt{x}} \cdot \frac{\sqrt{x}}{\sqrt{x}}$$

$$= \frac{1,000x \sqrt{x}}{x}$$

$$= 1,000 \sqrt{x}$$

THE SUPERIOR BRAIN OF WOMEN AND CALCULUS

Differentiate:

$$R(x) \quad = \quad 1{,}000\sqrt{x}$$

$$R'(x) \quad = \quad \frac{500}{\sqrt{x}}$$

$$R\,(100) \quad = \quad \frac{500}{\sqrt{100}}$$

$$= \quad 50$$

Answer: The approximate revenue from selling the first 10 gadgets is $50.

Solve the marginal profit, where P stands for profit and P(x) is revenue minus cost.

$$P\,(x) = R(x) \qquad\qquad -\,C(x)$$

$$= 1{,}000\sqrt{x} - (10x + 100\sqrt{x} + 10{,}000)$$
$$= \qquad -10x + 900\sqrt{x} - 10{,}000$$

Differentiate:

$$P\,(x) \quad = \quad 10x + 900\sqrt{x} - 10{,}000$$

$$P'(x) \quad = \quad -10 + \frac{450}{\sqrt{x}}$$

$$P'\,(100) \quad = \quad -10 + \frac{450}{\sqrt{100}}$$

$$= \quad -10 + 45$$

$$= \quad 35$$

Answer: The approximate profit for selling the first 10 gadgets is **$35**.

You get the same answer using the **shortcut:**

$$P'(x) \;=\; R(x) \;-\; C(x)$$

$$P'(100) = R'(100) \;-\; C'(100)$$

$$= \;\; 50 \;\; - \;\; 15$$

$$= \;\; 35$$

1. **Determine** the *maximum profit:*

$$P'(x) \;=\; -10 + \frac{450}{\sqrt{x}}$$

$$0 \;=\; 10 + \frac{450}{\sqrt{x}}$$

$$10 \;=\; \frac{450}{\sqrt{x}}$$

$$10\sqrt{x} \;=\; 450$$

$$\sqrt{x} \;=\; 45$$

$$\sqrt{x} \;=\; 2,025$$

2. **Plug** 2,025 into P(x) to find the maximum profit:

$$P(x) \;=\; -10x \;\; + 900\sqrt{x} \;\; - 10,000$$

$$P(2,025) = -10 \cdot 2,025 + 900\sqrt{2,025} - 10,000$$

$$= -20,250 \;+\; 900 \cdot 45 \;\; - 10,000$$

$$= \; 10,250$$

Answer: The maximum profit is $10,250.

Determine the profit-maximizing price by plugging the number sold into the demand function.

$$p = \frac{1{,}000}{\sqrt{x}}$$

$$p = \frac{1000}{\sqrt{2{,}025}}$$

$$= \frac{1{,}000}{45}$$

$$= 22.22$$

Therefore, the maximum profit of $10,250 happens when the price is set at $22.22. At this price, 2,025 gadgets will be sold.

Integration

As stated before, **Integration,** in calculus is just a fancy denomination for **addition**. It is merely a process of taking an undetermined area, breaking it up into infinitesimal pieces whose area is determined, and finally *adding them up* to get the *area of the whole.*

This process is basically getting a **finite sum** through *adding up an infinite number of numbers.* This process is paradoxically similar to the physical process of walking a certain distance. If we try to cover this distance through an infinite number of steps, each taking one second, we will never get to our destination. On the contrary, if we cover the same distance knowing that it will take us one second to reach our destination—by maintaining a *constant speed,* not stopping or slowing down at the end of each step—we will still take an infinite number of steps, but we will get to our destination in one second!

As the cornerstone of *differentiation* is comprised of two ideas—calculating rates and slope of a curve—the cornerstone of *integration* is also comprised of two ideas: adding up small pieces/ amounts and area under a curve.

To summarize, *integration* is calculating the area or volume of each thin strip from top to bottom and then adding up all the areas or volumes. This can be accomplished by cutting up an *infinite* number of *infinitely* thin slices to get an exact area or volume.

This process of *adding up* has an elegant *integration* symbol, which is simply an elongated $S, \int$.

$$\int_{bottom}^{top} dB^* = B \qquad \text{or} \qquad \int_a^b f(x) \ dx \ ^*$$

$$\int_{t=0}^{t=20} \text{little piece of distance} \ ^* \qquad \int_a^b \text{little piece of something}$$

*   *dB* means a little bit of a ball, and that by adding up all the small pieces from bottom to top, the result is the volume of the whole ball. On the other side, **f(x) dx** means to add up the little bits of area of all narrow rectangular strips between a and b under the curve f(x) in a graph to arrive at the total area under a curve. On the **little piece of distance** means to add up the little pieces of distance traveled between 0 and 20 seconds to obtain the total distance traveled during that interval.

The summation as an adding up of the areas of thin rectangular strips under a curve can be representing little bits of *distance*, *volume*, *energy*, or just an *area*. In a volume problem, each thin rectangle has a width measured in inches and a height measured in square inches; thus, the y-axis is labeled in *square* inches. Its area represents inches times square inches, or in other words, cubic inches of volume.

When integration is used to calculate area:

Area *below* the x-axis   =   **Negative area**

Total area between **a** and **b**

for a curve given by integral:

$$\int_b^a f(x)\,dx \qquad\qquad = \qquad \textbf{Net area*}$$

*   Total area *below the x-axis* and *above the curve* is subtracted from the total area *above the x-axis* and *below the curve.*

The Two Ways of Integration

There are two ways to calculate the area under a function:

1) Use the *most important discovery in mathematics about definition of integration* — **Anti differentiation (Integration is *differentiation in reverse*).**

2) Use the formal definition of integration — **the Limit of Riemann Sums.** (A short description of the Limit of Riemann Sums is given in appendix 1 of this book, since the author find it unnecessary to have a full description given its very laborious method.)

Anti Differentiation
Integration-Differentiation In Reverse

Instead of using the hard way to calculate the area under a function with the formal definition of integration under *The Limit of Riemann Sums*, we will use only *Anti differentiation.*

Anti differentiation is just differentiation backwards. For example,

Function	Derivative	Anti derivative
$\sin x$	$\cos x$	$\sin x$
x^2	$3x^2$	x^3
$x + 10$	$3x^2$	$x^3 + C^*$

*   where C is any number.

Indefinite Integral of a Function
Family of All *Antiderivatives* of a Function

$$\int f(x)\, dx$$

Example 1: The derivative of x^2 is $3x^2$; therefore, we say that the *Indefinite integral* of $3x^2$ is $x^3 + C$.

Family of all *Antiderivatives*

$$\int 3x^2 \, dx^3 = x + C$$

$\int$ = Indefinite integral/sum

$3x^2$ = Derivative

dx = All little bits of x

$X^3 + C =$ Indefinite Integral = the whole of x/family of all *anti-derivatives.*

Example 2: What is the *integral* of x^2? $\dfrac{1}{3}x + C$

Functions That Are Antiderivatives of the Same Function and with the Same *Derivative*

Function	Same Derivative	Same Antiderivative
$x^3 + 6$	$3x^2$	$x^3 + C*$
$x^3 + 2$	$3x^2$	$x^3 + C$
$x^3 - 6$	$3x^2$	$x^3 + C$

* x + C is the *family of all antiderivatives of* $3x^2$. Therefore, the *family of all antiderivatives and indefinite integral of* $3x^2$ is $x^3 + C$.

Area Function
Theorem of Calculus (First Version)

$$Af\,(\mathbf{x}) = \int_{s}^{x} \mathbf{f\,(t)\,dt}$$

Given an *area function Af* that sweeps out area under *f(t)*, the rate at which the area is being swept out is equal to the *height* of the original function. Therefore, because the *rate* is the *derivative*, the derivative of the area function **equals** the original function *f(x)*.

Derivative of *Af* (x): $\dfrac{\mathbf{d}}{\mathbf{dx}}\ Af\,(\mathbf{x}) = f(x)$

Because $Af\,(\mathbf{x}) = \int_{s}^{x} f\,(\mathbf{t})\,\mathbf{dt}$, the above equation can also be written as follows:

$$\frac{\mathbf{d}}{\mathbf{dx}}\ \int_{s}^{x} f\,(\mathbf{t})\,\mathbf{dt} = f(x)$$

Note:

Subscript f indicates that *Af (x)* is the *area function* with the *input variable x* for the particular **curve** *f(t)* that sweeps out area under it.

Point s indicates that "s" is for the starting point.

Input variable t is used in *f(t)* instead of x, because x is already taken—*x is the input variable of Af(x)*.

dt is an infinitesimally small increment along the *t-axis*.

Example: Area under f (t) = *10* between **3** and **x.** This area is swept out by the *moving vertical line at x.*

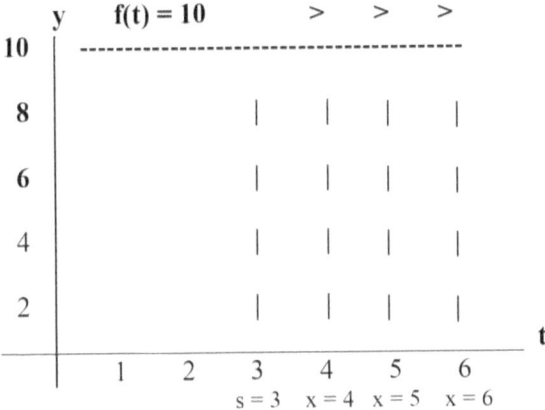

Derivative:

Theorem of calculus: $Af(x) = \int_s^x f(t)\,dt$ $\dfrac{d}{dx}\ Af(x) = f(x)$

Function: $f(t) = 10$ $Af(x) = \int_3^x 10\,dt\ ^*$ $\dfrac{d}{dx}\ Af(x) = 10\ ^{**}$

(horizontal line at y =*10*)

Sweep out area at s = 3 (from 3 to 4)

* From **3** to **4** is *10* because in dragging the line from **3** to **4**, a *rectangle unit* is swept out with a *width* of **1**, *height* of *10*, and *area* of **1** times 10 = *10*. And from **4** to **5** is 20; a *rectangle* with a *width* of **2**, *height* of *10*, *and area of **2** x *10* = 20. From **5** to **6** is 30; a *rectangle with a width* of **3**, *height* of *10*, *and area of **3** x *10* = 30*.

So $Af(4) = 10$, $Af(5) = 20$ and, $Af(6) = 30$, etc.

** The antiderivative of 10 is 10x + C, where the value of C depends on where the area starts to be swept out.

Remember: A *derivative* is a *rate*, which in this case is *10,* that equals the *height of the original function.* Therefore, the <u>rate of area</u> being swept out <u>under a curve</u> by an *area function* at a given *x-value* is **equal** to the <u>height of the curve</u> at that *x—value.*

Rate of growth: With each one *unit* increase in **x,** the *area function Af* goes **up** *10,* so the *rate of growth* is **ten square units per second** if the line is dragged across at a rate of *one unit per second.*

Example: Area under *f* (t) between **2** and **x.** This area is swept out by the *moving vertical line* at **x.**

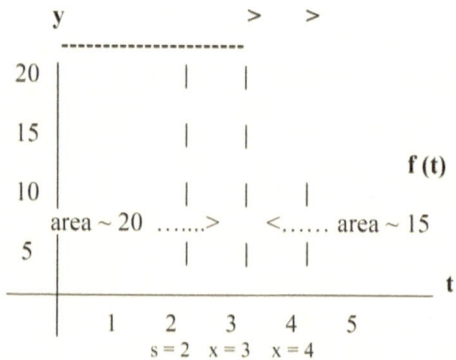

Note: *Af* (3) is about 20; because the area swept out between **2** and **3** has a *width* of **1** and the *curved top of the rectangle* has an approximate *height* of 20, its *rate of growth* is about **20** *square units per second.* In the interval from x = **3** to x = **4,** its *rate of growth* is about **15.**

Example: Find the exact area under $f(x) = x^2 + 1$ between 0 and **3.** The area function for sweeping out area under $x^2 + 1$ is:

$$Af\ (x)\ =\ \int_{s}^{x}(t^2\ +1)\ dt$$

Area Function	Function Form	Derivative	Anti derivative
$Af(x) = \int_s^x (t+1)\,dt$	$\frac{1}{3}x^s + x$	$\frac{d}{dx} Af(x) = x^2 + 1$	$\frac{1}{3}x^s + x + C$
$Af(x) = f(t)^* \, dt^*$	Af	$\frac{d}{dx} Af(x) = f(x)$	$Af + C$

* Using the Theorem for Calculus,

when s = 0, C = 0; therefore,

$$Af(x) = \int_0^x (t^2 + 1)\,dt = \frac{1}{3}x^3 + x \text{ (with a C value of zero).}$$

The area swept out from 0 to **3** is simply *Af* (3). This same problem can be done the *hard way* by computing the *limit of a Riemann sum*.

$$Af(x) = \frac{1}{3}x^3 + x$$

$$Af(3) = \frac{1}{3} \cdot 3^3 + 3 \text{ *}$$

$$= 9 + 3$$

$$= 12$$

* $3^3 = 3 \times 3 \times 3 = 27 \div 3 = 9$

Note: Finding a parabola ($f(x) = x^2 + 1$) between **2** and **3** is easy, simply by subtracting the area between 0 and 2 from the area between 0 and **3**. For example,

Area between 0 and 3	=	12
Area between 0 and 2	=	$4\,\frac{2}{3}$ *
Area between 2 and 3	=	$7\,\frac{1}{3}$

* Area between 0 and 2 $= Af(2) = \frac{1}{3}\,2^{3**} + 2 = 2\,\frac{2}{3} + 4\,\frac{2}{3}$

** $\frac{1}{3} \cdot \frac{8}{1} = \frac{8}{3} = 2\,\frac{2}{3}$

Theorem of Calculus (Second Version)

$$\int_a^b f(x)\,dx = F(b) - F(a)$$

Let *F* be any *antiderivative* of the function *f*.

The advantage of the second version theorem is that *we just need to find any antiderivative, F(x), of our function and do the subtraction: F(b)—F(a).* The simplest antiderivative to use is the derivative where C = 0.

Example: Use the second version theorem to find the area under Parabola function from **2** to **3**: $x^2 + 1$

Antiderivative: $F(x) = \frac{1}{3}x^3 + x$ with a C value of zero

Area between 0 and 2: $Af(2) = \frac{1}{3}\,2^3 + 2 = 4\,\frac{2}{3}$

Theorem:

$$\int_a^b f(x)\ dx\ =\ F(b)\ -\ F(a)$$

$$\int_2^3 (x^2+1)\ dx\ =\ F(3)\ -\ F(2)$$

$$=\ 1\cdot 3^3 + 3\ -\ (\frac{1}{3}\cdot 2^3 + 2)$$

$$=\ 12\ -\ 4\frac{2}{3}$$

$$=\ 7\frac{1}{3}$$

Example: Find the area between x = 3 and x = 5 under

Function: $\qquad f(x)\ =\ e^{x}$ *

Derivative: $\qquad e^{x}$

Antiderivative: $\qquad e^{x}$

* $\quad e \approx 2.7182818\ldots$

Theorem of Calculus $\qquad \int_a^b f(x)\ dx\ =\ F(b) - F(a)$

$$\int_3^5 e^x\ dx\ =\ F(e)-F(e)$$

$$\approx\ 148.4\ -\ 20.1$$

$$\approx\ 128.3$$

Sigma—The Summation Notation

Sigma (the eighteenth letter of the Greek alphabet) notation is used for adding up long series of numbers like the rectangle areas in a left, right, or midpoint sum. For example,

$$\sum_{i=1}^{100} 5i\ =\ 5+10+15+20+25+\ldots+490+495+500$$

Note: This notation is simply saying in the *bottom*, $i = 1$, to replace i and plug 1 in $5i$:

First multiply 5 by 1 = **5**
Then multiply 5 by 2 = **10**
Then multiply 5 by 3 = **15**
Then multiply 5 by 4 = **20**
Then multiply 5 by 5 = **25** and so on, up to 5 x 100 = **500**

Notice that the number 100 is indicated on top of the sigma notation. Another example using *powers:*

$$\sum_{k=10}^{30} k = 10 + 11 + 12 + ... + 29 + 30$$

Definite Integral vs Riemann Sums

The Definite Integral.

$$\int_a^b f(x)\, dx$$

The definite integral from **a** to **b** is the number to which all *Riemann sums* tend as the number of rectangles approaches infinity and as the width of all rectangles tend to approach zero.

$$\int_a^b f(x)\, dx = \lim_{n \to \infty} \sum_{i=1}^{n} f(c_i)\, \Delta x_i$$

where Δx_i is the width of the ith rectangle and c_i is the x-coordinate of the point where the ith rectangle touches $f(x)$.

Five Definite Integral Rules

1. $\int_a^a f(x)dx$ $= 0$

2. $\int_b^a f(x)dx$ $= -\int_a^b f(x)dx$

3. $\int_a^b f(x)dx$ $= \int_a^c f(x)dx + \int_c^b f(x)dx$

4. $\int_a^b kf(x)dx$ $= k\int_a^b f(x)dx$

5. $\int_a^b [f(x) + g(x)]dx = \int_a^b f(x)dx + \int_a^b g(x)dx$

Visualization of Three Area Functions

$f(t) = 10$: $Af(x)$ starting at 0 point (in which C = 0)

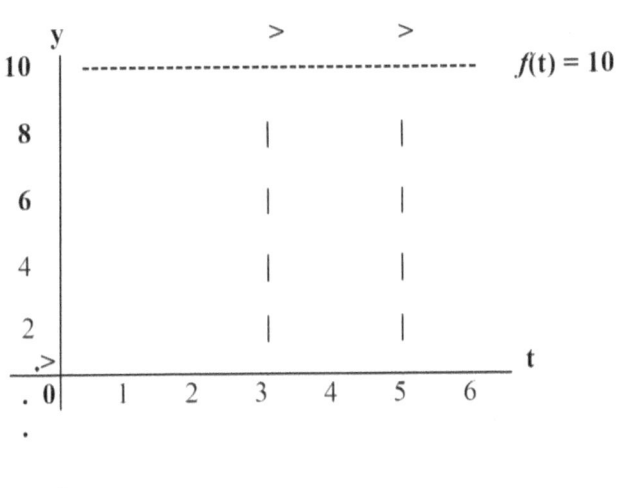

s = 0

$f(t) = 10$: Bf (x) starting at **-2** (in which **C = 20**)

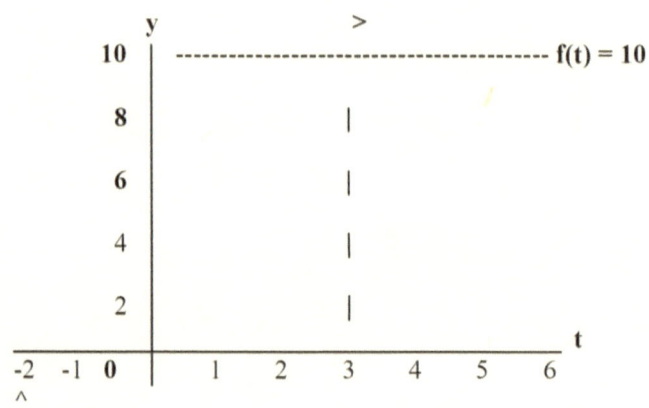

s = **-2**

f(t) = 10: $Bf(x)$ starting from **5** to **8** (in which C = **30**)

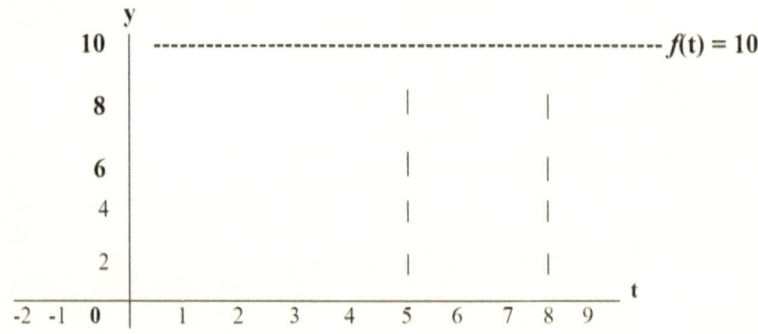

Theorem: $\int_a^b f(x)\ dx \ = \ F(b) - F(a)$

$$\int_5^8 10 \quad dx \ = \ Af(8) - Af(5)$$

$$=$$

$$10 \cdot 8 \ + \ 20 \ - \ 10 \cdot 5 \ + \ 20 \ ^{*}$$

$$= \quad 80 + 20 - (50 + 20)$$
$$= \quad 80 + 20 - 50 - 20$$

(Cancel +20 and -20) $= \quad 80 - 50$
$$= \quad 30$$

*   100 is the *area of the rectangle from* -2 to 8 (width from -2 to 5 is 70)

Differentiation and Integration Connection

Function

$$f(x) = x^2 + x$$

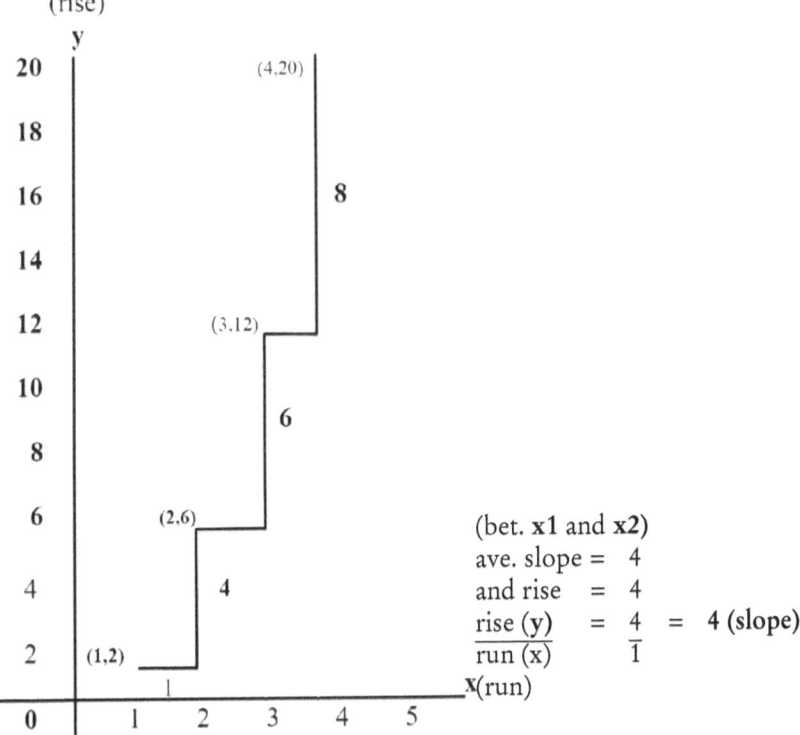

(rise)

(bet. **x1** and **x2**)
ave. slope = 4
and rise = 4
$\dfrac{\text{rise (y)}}{\text{run (x)}}$ = $\dfrac{4}{1}$ = 4 (slope)

Note on the points: (1,2), 1 pertains to **x** (run), and **2** pertains to **y** (rise).

Derivative

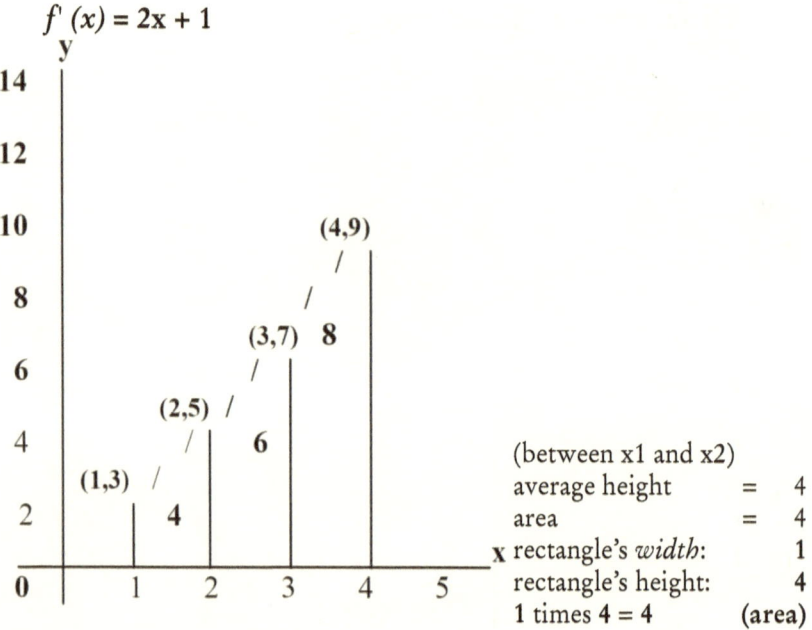

$f'(x) = 2x + 1$

These two graphs are a synthesis of countless symbols and equations, giving the essence of integration. It gives a clear picture of the connection between **4, 6,** and **8** on the graph of f, which are the amounts of *rise* on the **curve**, and **4, 6,** and **8** on the graph of f', which are the *areas* of the **trapezoids** under f'. It also shows how the second version of the fundamental theorem operates because it depicts that the *area* under 2x + 1 between **1** and **4** equals the *total rise* on f, between (1,2) and (4,20). Notice that by emphasizing that:

$x^2 + 1$ (called F) is an *antiderivative of* **2x + 1,** the area equation is written:

$$\int_1^4 f(x)\, dx \quad = \quad F(4) - F(1)$$

And by emphasizing that:

2x + 1 (called *f*) is the *derivative of* $x^2 + 1$, the area equation is written:

$$\int_1^4 f(x)\ dx\ =\ f(4) - f(1)$$

In synthesis, that is what the second version of the fundamental theorem says:

$$\int_a^b f'(x)\ dx\ =\ f(b) - f(a)$$

area = rise

Notice that *total area under f'* between **1** to **4** on the x-axis (4 +6+8) = **18**

Corresponds to:

Total rise under f between **1** to **4** on the x-axis (4+6+8) = **18**

x + 1 is an *antiderivative of* **2x + 1**, the area equation is written:

$$\int_1^4 f(x)\ dx\ =\ F(4) - F(1)$$

Example: What is the area between **2** and **3** under $x^2 + 10$:

Function: $\int_2^3 (x^2 + 10)\ dx$ $= F(3) - F(2)$

Derivative: $= \left[\dfrac{1x^3}{3} + 10\,x\right]^3$

Substitute and solve:

$$= \frac{1^2}{3} \cdot 3 + 10.3 - \left(\frac{1}{3} \cdot 2^3 + 10 \cdot 2 \right)$$

F(2):

$$= \frac{1^2}{3} \cdot 3 + 10 \cdot 3 - 22 \tfrac{2}{3}$$

Answer: The opposite of F(2), which is **-22** 2/3 for C-value, is the area function for: **2**

$$x^2 + 1$$

Seven Techniques to Find Antiderivatives

First Technique

           Reverse Rules for Antiderivatives Method

Second Technique

           Trying and Checking Method

Third Technique

           Substitution Method

Fourth Technique

           Integration by Parts Method

Fifth Technique

           Trigonometric Integrals Method

Sixth Technique

           Trigonometric Substitution Method

Seventh Technique

           Partial Fractions Method

All of these techniques, except for the first technique, are not given in full detail in this book, but a brief description is found in appendix 2. Fortunately, all of these techniques can be replaced by using a scientific calculator to find them quickly and efficiently.

First Technique
Reverse Rules for Antiderivatives Method

The reverse of the derivative rule is very simple. It tells us that if a certain function has a *derivative*, the function becomes automatically the *antiderivative* of that particular *derivative*, For example,

Function:	Derivative:	Antiderivative:
sinx	cosx	sinx + C *

* All functions of the form sinx + C are *antiderivatives* of cosx. In symbols, we have:

Derivative: $\dfrac{d}{dx} \sin x = \cos x$

Antiderivatives of cosx: $\int \cos x\, dx = \sin x + C$

14 Basic Antiderivative Formulas

1) $\int \quad dx = x + C$

2) $\int a^2 \quad dx = \dfrac{1}{\ln a} \quad a^x + C$

3) $\int e^x \quad dx = e^x + C$

4) $\int \cos x \quad dx = \sin \quad x + C$

5) $\int \sin x \quad dx = -\cos \quad x + C$

6) $\int \sec^2 x \quad dx = \tan \quad x + C$

7) $\int \csc x \cot x \; dx \;\; = \; -\csc \;\;\; x + C$

8) $\int \sec x \tan x \; dx \;\; = \; \sec \;\;\; x + C$

9) $\int \csc^2 x \;\;\;\; dx \;\; = \; -\cot \;\;\; x + C$

10) $\int x^n \;\;\;\;\; dx \;\; = \;\; \dfrac{x^{n+1}}{n+x} + C$

11) $\int \;\;\; \dfrac{dx}{x} \;\; = \; \ln \;\;\; |x| + C$

12) $\int \;\;\; \dfrac{dx}{a^2 + x^2} \;\; = \; \dfrac{1}{a} \arctan \;\; \dfrac{x}{a} + C$

13) $\int \;\;\; \dfrac{dx}{\sqrt{a^2 - x^2}} \;\; = \; \arcsin \;\; \dfrac{x}{a} + C$

14) $\int \;\;\; \dfrac{dx}{x\sqrt{x^2 - a^2}} \;\; = \; \dfrac{1}{a} \text{arcsec} \;\; \dfrac{|x|}{a} + C$

*   Remember the power rule—derivative of $5x^4$ is $20x^3$. If we have a coefficient, in this case **5,** *first* we multiply the power by the coefficient: 4 x 5 = 20, then we reduce the power so 1: 4 is reduced to 3.

More examples with power rule:

Function:	Derivative:	Anti derivative:
x^3	$\dfrac{d}{dx} x = 3x$	$\int 3x^2 \; dx = x^3 + C$
$5x^4$	$\dfrac{d}{dx} 5x^4 = 20x^3$	$\int 20x^3 \; dx = 5x^4 + C$

For *reversing the power rule:*

1. Increase the power by one: $\qquad\qquad$ $20x^3 \rightarrow 20x^4$

2. Divide by the new power and simplify: $\qquad$ $\dfrac{20}{4}x^4 = 5x^4$

Therefore, $\qquad\qquad$ $\displaystyle\int 20x^3 \, dx = 5x^4 + C$

This method is very useful to test antiderivatives. If we can get back to the original function, we know the antiderivative is correct. For example,

Original function: $\qquad\qquad\qquad\qquad\qquad$ $5x^4$

Derivative: $\qquad\qquad\qquad\qquad\qquad\qquad$ $20x^3$

Increase power by one: $\qquad\qquad\qquad\qquad$ $20x^4$

Divide by new power and simplify: $\qquad$ $\dfrac{20x^4}{4} = 5x^4$

Once we found the correct antiderivative, we are able to find the area under

$20x^3$ between **1** and **2**:

$$\int 20x^3 \, dx \quad = \quad 5x^4 + C$$

$$\int 20x \, dx \quad = \quad \left[5x4\right]_1^2$$

$$= \quad 5 \cdot 2^2 - 5 \cdot 1^4$$

$$= \quad 80 - 5$$

$$= \quad 75$$

Mean Value Theorem
for Integrals and Average Value

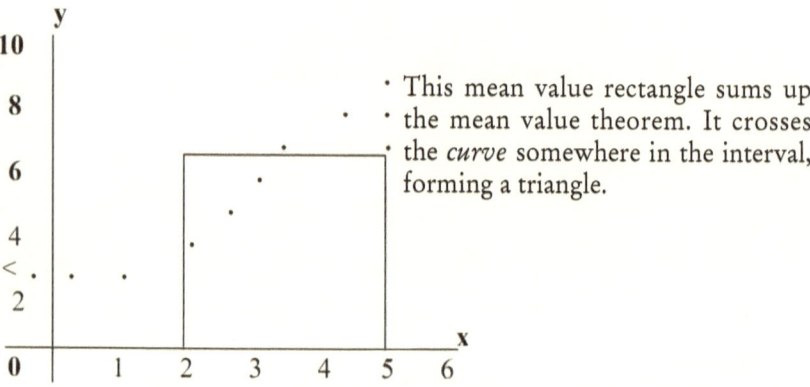

This mean value rectangle sums up the mean value theorem. It crosses the *curve* somewhere in the interval, forming a triangle.

If $f(x)$ is a continuous function on the closed interval [a,b], then there exists a number c in the closed interval, such that:

$$\int_a^b f(x)\,dx = f(c) \cdot (b-a)$$

$$area = height \bullet base$$

Basically, the theorem just corroborates the existence of the mean value rectangle.

The area under the curve, which is the same as the area of the mean value rectangle, equals *height* times *base*. Therefore, if we divide the *area* by its *base* we get its *height,* and this *height* of the mean value rectangle is the **average value** of a function $f(x)$ over a closed interval [a,b].

$$\frac{1}{b-a} \int_a^b f(x)\,dx$$

In synthesis: *function value* (*area* under curve) = *height* = **speed**

$$base = \textbf{time}$$

$$area = speed \cdot time = \textbf{distance}$$

$$(height)\ (base)$$

Therefore,

$$average\ \textbf{speed} = \frac{total\ \textbf{distance}\ (area)}{total\ \textbf{time}\ (base)}$$

$$(height) \qquad\qquad$$

Examples:

What is the *average speed* between t = 9 seconds and t = 16 seconds, when speed is in *feet per second* given by the function:

$$f\ (t)\ =\ 30\ \sqrt{t}$$

Average value
$$\frac{1}{b-a}\ \int_{a}^{b} f\ (x)\ dx$$

Average speed:
$$\frac{1}{16-9}\ \int_{9}^{16} 30\ (t)\ dt$$

$$\frac{1}{16-9}\ \int_{9}^{16} 30\ (t)\ tx$$

$$=\quad 30\left[\frac{2}{3}\ t^{3/2}\right]_{9}^{16}$$

$$=\quad \left(\frac{128}{3}\ -\ \frac{54}{3}\right)$$

$$=\quad 740$$

* $\dfrac{30}{1} \times \dfrac{2}{3} = \dfrac{60}{3} = \dfrac{20}{3} = 6 \times 16 = \dfrac{96}{3} = 32$

$32 = \dfrac{32}{1} \quad \dfrac{32}{1} + \dfrac{96}{3} = \dfrac{32}{3} + \dfrac{96}{3} = \dfrac{128}{3}$

$\dfrac{30}{1} \times \dfrac{2}{3} = \dfrac{60}{3} = \dfrac{20}{3} = 6 \times 9 = \dfrac{54}{3}$

average speed $= \dfrac{total\ \mathbf{distance}}{total\ \mathbf{time}} = \dfrac{740\ \text{feet}}{7\ \text{seconds}}$ $(16 - 9 = 7)$

Answer: ≈ 105.7 feet per second

Differentiation/Integration Connection

<u>Mean Value Theorem</u> for Derivatives

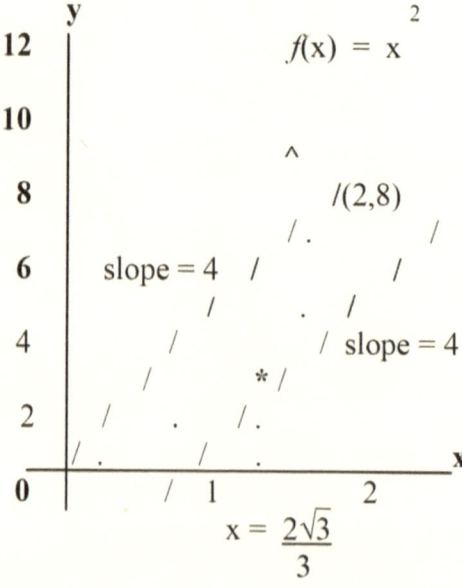

Mean Value Theorem
for Integrals

$$f'(x) = 3x$$

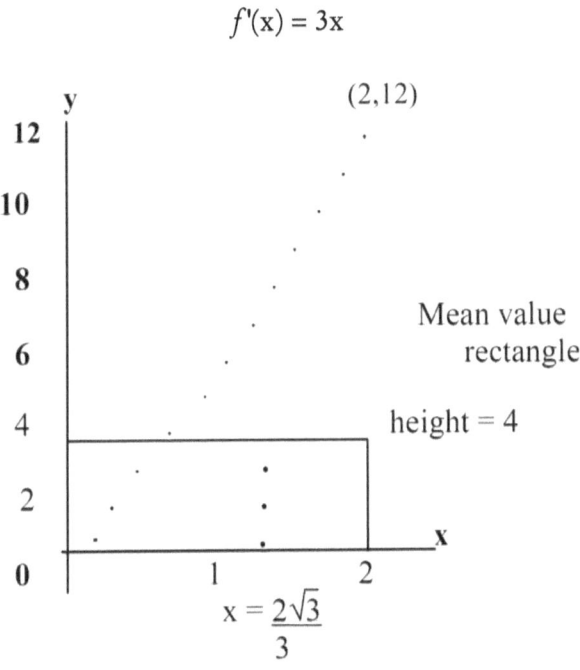

$$x = \frac{2\sqrt{3}}{3}$$

At $x = \dfrac{2\sqrt{3}}{3}$

—*Slope* = 4
   (average slope of f bet0 and 2)

—*Least slope* of f = 0

—*Greatest slope* of f = 12

—*Total Rise along* f = 8
   (from 0 to 2)

At $x = \dfrac{2\sqrt{3}}{3}$

—*Height* = 4
   (average height of f bet. 0 and 2)

—*Least height* of f' = 0

—*Greatest height* of f' = 12

—*Total Area under* f' = 8
   (from 0 to 2)

16 Different Problems

Finding Areas

Problem 1

Find the area of a rectangle on a curve between *y-coordinate*:

(height) $y = 2 - x^2$ and (base) $y = \dfrac{1}{2} x$ from x = 0 to x = 1.

1. **Get the rectangle's height** by *substracting* its base, which is the infinitesimal **dx**, from height:

$$\text{area} = \text{height} \cdot \text{base}$$

Rectangle area $\qquad = \left[\left(2 - x^2\right) - \dfrac{1}{2} x \right] \qquad \mathbf{dx}$

2. **Integrate** by *adding* all rectangles from 0 to 1:

$$\int_0^1 \left[\left(2 - x^2\right) - \dfrac{1}{2} x \right] \qquad \mathbf{dx}$$

$$= \left[\left(2x - \dfrac{1}{3} x \right)^3 - \dfrac{1}{4} x^2 \right]_0^1 \qquad \text{(reverse power rule)}$$

$$= \left(2 - \dfrac{1}{3} - \dfrac{1}{4} \right) - 0 - 0 - 0 \, \text{(substitute)}^*$$

Answer: $\quad = \dfrac{17}{12}$ square units

* $\quad 2 - \dfrac{1}{3} - \dfrac{1}{4} = \dfrac{24}{12} - \dfrac{4}{12} - \dfrac{3}{12} = \dfrac{24}{12} - \dfrac{7}{12} = \dfrac{17}{12}$

Problem 2

Find area between $3\sqrt{x}$ and x from x = 0 to x = 2. There are two curves crossing each other and dividing the area into two separate regions, one from 0 to 1 and another from 1 to 2. The curves cross at (1,1).

First find area of the region on left rectangle on a curve between *y-coordinate* from 0 to 1:

(height) $y = \sqrt{x}$ and (base) $y = x^3$ from x = 0 to x = 1.

1. **Get the rectangle's height** by *substracting* its base, which is the infinitesimal **dx,** from height and **figure the area on the left:**

    $3\sqrt{x}$ is above x^3

    area = height · base

    Rectangle area = $(3\sqrt{x} - x^3)$ dx

2. **Integrate** by *adding* all rectangles from 0 to 1:

    $$\int_0^1 (3\sqrt{x} - x^3)\, dx$$

    $$= \left[\frac{3}{4} x^{4/3} - \frac{1}{4}x^4 \right]_0^1 \quad \text{(reverse power rule)}$$

    $$= (\frac{3}{4} - \frac{1}{4}) - (0 - 0) \quad \text{(substitute)}^*$$

    $$= \frac{1}{2}$$

*   substitute $x = 1$: $\quad \frac{3}{4} \; x^{4/3} - \frac{1}{4}x = \frac{3}{4} \; 1^{4/3} - \frac{1}{4} \; 1^4 = \frac{3}{4} - \frac{1}{4}$

substitute $x = 0$: $\quad \frac{3}{4} \; x^4 - 1x^4 = \frac{3}{4} \; 0^{4/3} - \frac{1}{4} \; 0^4 = 0 - 0$

Find area of the region on right rectangle on a curve between *y coordinate* **from 1 to 2:**

(height) $y = x^3$ and (base) $y = 3\sqrt{x}$ from $x = 1$ to $x = 2$.

1. **Get the rectangle's height** by *substracting* its base, which is the infinitesimal **dx**, from height, and **figure the area on the right:**

$$x^3 - 3\sqrt{x}$$

$$\text{area} = \text{height} \cdot \text{base}$$

$$\textbf{Rectangle area} = \left(x^3 - 3\sqrt{x}\right) \, \textbf{dx}$$

2. **Integrate** by *adding* all rectangles from 0 to 1.

$$\int_{1}^{2} (x^3 - 3\sqrt{x}) \, dx$$

$$= \left[\frac{1}{4} \, x^4 - \frac{3}{4} \, x^{4/3}\right]_{1}^{2} \text{(reverse power rule)}$$

3. **Substitute*** $\quad = \left(4 - \frac{3}{2} \; 3\sqrt{2}\right) - \left(\frac{1}{4} - \frac{3}{4}\right)$

Answer: $\quad = 4.5 - 1.5 \; 3\sqrt{2}$

* $\frac{1}{4} - \frac{3}{4} = \frac{2}{4} = \frac{1}{2}$ equals in decimals $+ \; 0.5(4) = 4.5 - \frac{3}{2}\left(\frac{3}{2} = \frac{2}{3} = 1.5\right)$

Note: The height of a representative rectangle is *always* its top minus its bottom, whether the number is negative or positive. For example,

Rectangle that goes from: Height:

30 *up* to 40			40	-	30	or	10
-5 *up* to 10			10	-	(-5)	or	15
-25 *up* to -20			-20	-	(-25)	or	5

Finding Volumes

Problem 3

Find the volume of the solid between x = 2 and x = 3, which is generated by rotating the curve:

$$y = e^x \text{ about the x-axis}$$

1. **Determine cross section area** (cross section is a circle with a radius of e^x):

Circle's area: $A = \pi r^2$

Cross section Circle's area: $A = \pi(e^x)^2 = A = \pi e^{2x}$

2. **Get the volume** of an infinitely small dx:

Volume 1 = area • thickness

Volume = πe^{2x} • dx

3. **Integrate** by adding up the volumes from 2 to 3:

$$= \int_2^3 \pi e^{2x} \ dx$$

$$= \pi \ \int_2^3 e \ dx^{2x}$$

$$= \frac{\pi}{2} \Big[\ e^{2x} \ \Big]_2^3$$

4. **Substitute** by multiplying: $= \dfrac{\pi}{2}^6 \left(e^4 \ - \ e \right)$

Answer: ≈ 548 *cubic unit*

* $\pi \approx 3.14, \ e \approx 2.71 \ \left[\dfrac{\pi}{2} (e^6 - e^4) \approx 1.57 \ (\ 2.71^6 - 2.71^4) \approx 548 \right]$

Problem 4

Find the volume of the solid between x = 2, x = 3, and $y = e^x$, which is generated by revolving the area about the y-axis.

1. **Determine the area** of a cylindrical shape:

Rectangle Area	=	length	·	width
Cylindrical Area:	=	$2\pi r$	·	h
		(circumference)		(height)
	=	$2nx$	·	e^x *

*   radius = x and height = e^x

(y = e^x, and *y-axis* is the vertical line that *always represents* the *height*).

2. **Get the volume** of an infinitely small dx:

$$\text{Volume} \quad = \quad \text{area} \quad \cdot \quad \text{thickness}$$

$$\text{Volume} \quad = \quad 2\pi x e^{x} \quad \cdot \quad dx$$

3. **Integrate** by adding up the volumes from 2 to 3:

$$= \quad \int_{2}^{3} 2\pi x e^{x} \; dx$$

$$= \quad 2\pi \int_{2}^{3} x e^{x} \; dx$$

$$= \quad 2\pi \left[x e^{x} - e^{x} \right]_{2}^{3}$$

4. **Substitute** by multiplying:

$$= 2\pi \left[3e^{3} - e^{3} - \left(2e^{2} - e^{2} \right) \right]$$

$$= 2\pi \qquad\qquad \left(2e^{3} - e^{2} \right)$$

Answer: ≈ 206 *cubic units*

* $\pi \approx 3.14, \; e \approx 2.71 \left[2\pi \left(2e^{3} - e^{2} \right) \approx 6.28 \left(2(2.71)^{3} - 2.71^{2} \right) \approx 206 \right]$

Problem 5

Find the volume of the solid between $y = x^{2}$ and $y = \sqrt{x}$.

1. **Determine the two curves' intersection**, which spans the interval on the x-axis from 0 to 1 and its cross section area.

    The cross section area can be defined as a circle with two radius in which the area of the circle minus the hole is

$$\text{Area} = \pi R^{2} - \pi r^{2}$$

Integration: $\int_a^b \left(\pi R^2 - \pi r^2\right)\ dx$

or: $\pi \int_a^b \left(\pi R^2 - \pi r^2\right)\ dx$

Outer bigger radius (R) is $\sqrt{x}$ and inner smaller radius (r) is x^2

Circle's area: $A = \pi r^2$

Cross section Circle's area: $A = \pi\left(\sqrt{x}\right)^2 - \pi\left(x^2\right)^2$

$$= \pi x - \pi x^4$$

2. **Get the volume** of an infinitely small dx:

Volume $=$ area $\cdot$ thickness

Volume: $= \left(\pi x - \pi x^4\right) dx$

3. **Integrate** by adding up the volumes from 0 to 1:

$$= \int_0^1 \left(\pi x - \pi x\right)^4\ dx$$

$$= \pi \int_0^1 \left(x - x\right)^4\ dx$$

4. **Substitute:***

$$= \pi\left[\left(\frac{1}{2} - \frac{1}{5}\right) - (0 - 0)\right]$$

$$= \pi\left(\frac{3}{10}\right)$$

$$= \frac{3\pi}{10}$$

Answer: $= $ **0.94 cubic units** *

*   LCD and cross multiplication $\dfrac{1}{2} - \dfrac{1}{5} = \dfrac{5}{10} - \dfrac{2}{10} = \dfrac{3}{10} \cdot (\pi = 3.14) \times 3 = 0.942$

Finding Lengths

To get the precise length of the curve, all hypotenuses along the curve between *start* and *finish* points need to be added up; the *length* of the legs of each infinitesimal triangle are **dx** and **dy**. Therefore, following the Pythagorean Theorem:

Length of the hypotenuse is   **Integration** (adding up all hypotenuses From **a** to **b** along the curve) is:

$$\sqrt{(dx)^2 + (dy)^2} \qquad \int_a^b \sqrt{(dx)^2 + (dy)^2}$$

To get the **Arc Length Formula**, *the* (dx) under the square root needs to be *factored out* and *simplified*, and then its *square root* **dx** brought outside the radical.

Factor out: Simplify:

$$(dx)^2 \left[1 + \frac{(dy)^2}{(dx)^2} \right] \qquad (dx)^2 \left[1 + \frac{(dy)^2}{dx} \right]$$

Arc Length Formula

$$\int_a^b \sqrt{1 + \frac{(dy)^2}{dx}} \; dx$$

Problem 6

Find the length along $y = (x-1)^{3/2}$ from x = 1 to x = 5.

1. **Derive** the function:

$$y = (x-1)^{3/2}$$
$$\frac{dy}{dx} = \frac{3}{2}(x-1)^{3/2}$$

2. **Integrate** using the arc length formula:

Arc Length Formula

$$\int_a^b \sqrt{1 + \frac{(dy)^2}{dx}}\ dx$$

$$= \int_1^5 \sqrt{1 + \left(\frac{3}{2}(x-1)^{1/2}\right)^2}\ dx$$

$$= \int_1^5 \sqrt{1 + \frac{9}{4}(x-1)}\ dx$$

$$= \int_1^5 \sqrt{\left(\frac{9x}{4} - \frac{5}{4}\right)^{1/2}}\ dx$$

3. Use the **reverse power rule**:

$$= \left[\frac{4}{9} \cdot \frac{2}{3}\left(\frac{9}{4}x - \frac{5}{4}\right)^{3/2}\right]_1^5\ dx$$

4. **Substitute:**

$$= \left[\frac{1}{27}(9x - 5)^{3/2}\right]_1^5\ dx$$

$$= \frac{1}{27}\left(\sqrt{40}\right)^3 - \frac{1}{27} \cdot 8$$

$$= \frac{8}{27}\left(\sqrt{10}\right)^3 - 1$$

Answer: $\approx$ **9.07 units**

Problem 7

What is the surface area between x = 1 and x = 2 of the surface generated by revolving:

$$y = x^3 \text{ about the x axis?}$$

Note: *Radius* equals x^3, the length of rectangle equals the circumference $2\pi r$, so its *circumference* becomes :

$$2\pi x^3$$

its *width* (hypotenuse):

$$\sqrt{1 + \left(\frac{dy}{dx}\right)^2} \quad dx$$

1. **Derive** the function:

$$y = x^3$$
$$\frac{dy}{dx} = 3x^2$$

2. **Integrate** using the arc length formula:

Arc Length Formula

$$\int_a^b \sqrt{1 + \left(\frac{dy}{dx}\right)^2} \quad dx$$

$$\int_1^2 2\pi x^3 \sqrt{1 + \left(3x^2\right)^2} \quad dx$$

$$= 2\pi \quad \int_1^2 x^3 \sqrt{1 + 9x^4} \quad dx$$

3. **Substitute** with **u** and **x. Use** power rule:

$(u = 1 + 9x^4, du = 36x^3 dx$. When $x = 1$, $u = 10$, when $x = 2$, $u = 145)$

$$= 2\pi \int_1^2 36x^3 \sqrt{1 + 9x^4} \; dx$$

$$= \frac{\pi}{18} \int_{10}^{145} u^{1/2} \; du$$

4. **Use** reverse power rule:

$$= \frac{\pi}{18} \left[\frac{2}{3} u^{3/2} \right]_{10}^{145}$$

$$= \frac{\pi}{18} \left(\frac{2}{3} \cdot 145^{3/2} - \frac{2}{3} \cdot 10^{3/2} \right)$$

Answer: $\approx$ 199.5 square units

Problem 8

What is the $\lim\limits_{x \to \infty} \dfrac{ln\,(x)}{x}$?

Use the L'Hopital's Rule and substitute:
(derivatives of $ln(x)$ is $\dfrac{1}{x}$, and x is 1)

$$\lim\limits_{x \to 0} \frac{f(x)}{g(x)} = \lim\limits_{x \to 0} \frac{f'(x)}{g'(x)}$$

$$\lim\limits_{x \to \infty} \frac{ln(x)}{x} = \lim\limits_{x \to \infty} \frac{\frac{1}{x}}{\frac{1}{1}} = \frac{\frac{1}{\infty}}{\frac{1}{1}} = \frac{0}{1} = 0$$

Problem 9

Evaluate $\lim\limits_{x \to 0} \dfrac{e^{3x} - 1}{x}$

Use the L'Hopital's Rule and substitute:

(derivatives of $e^{3x} - 1$ is $3e^{3x}$, and x is 1)

$$\lim\limits_{x \to c} \frac{f(x)}{g(x)} = \lim\limits_{x \to c} \frac{f'(x)}{g'(x)}$$

$$\lim\limits_{x \to 0} \frac{e^{3x} - 1}{x} = \lim\limits_{x \to 0} \frac{3e^{3x}}{1} = \frac{3 \cdot 1}{1} = 3$$

Problem 10

Evaluate $\lim\limits_{x \to \infty} \left(e^{-x} \ \sqrt{x} \right)$

Substitution equals $0 \cdot \infty$; therefore, it needs to be twitched:

$$\lim\limits_{x \to \infty} \left(e^{-x} \ \sqrt{x} \right) = \lim\limits_{x \to \infty} \left[\frac{\sqrt{x}}{e^{x}} \right]$$

Use the L'Hopital's Rule and substitute:

(derivatives of $\sqrt{x}$ is $\dfrac{1}{2\sqrt{x}}$, and e^{x} is e^{x})

$$\lim\limits_{x \to c} \frac{f(x)}{g(x)} = \lim\limits_{x \to c} \frac{f'(x)}{g'(x)}$$

$$\lim\limits_{x \to \infty} \frac{\sqrt{x}}{e^{x}} = \lim\limits_{x \to \infty} \frac{\dfrac{1}{2\sqrt{x}}}{e^{x}} = \frac{\dfrac{1}{2\sqrt{\infty}}}{e^{\infty}} = \frac{\dfrac{1}{\infty}}{\infty} = \frac{0}{\infty} = 0$$

Problem 11

What is the area under $y = \dfrac{1}{x^2}$ from 0 to 1?

1. Turn *definite integral* into a *limit*:

$$\int_0^1 \frac{1}{x^2}\,dx = \lim_{c\to 0}+ \int_c^1 \frac{1}{x^2}\,dx$$

2. Use reverse power rule and substitute:

$$= \lim_{c\to 0^+} \left[\frac{1}{x}\right]_c^1$$

$$= \lim_{x\to 0^+} \left((-1)-\left(-\frac{1}{c}\right)\right)$$

$$= -1-(-\infty)$$

Answer: $= \infty$

Note: (-) (-) = (+) and this function (curve) goes up to *infinity* and its area is also *infinite*.

Comparison of Infinite and Finite Areas

Improper integrals can have *infinite* or *finite* areas, even though their *functions* (curves) go to *infinity* at x = 0. For example,

$$\int_0^1 \frac{1}{x^2}\,dx \qquad\qquad \int_0^1 \frac{1}{3\sqrt{x}}\,dx$$

Infinite Area $= \infty$ *Finite* Area $= \dfrac{3}{2}$

Note that in both *functions* (curves) the **definite integral**

$$\int_0^1 \quad \text{is exactly the same:} \quad \int_0^1$$

Problem 12

What is the area under $y = \dfrac{1}{3\sqrt{x}}$ from 0 to 1?

1. Turn *definite integral* into a *limit*:

$$\int_0^1 \frac{1}{3\sqrt{x}}\, dx = \lim_{c \to 0^+} \int_c^1 \frac{1}{3\sqrt{x}}\, dx$$

2. Use reverse power rule and substitute:

$$= \lim_{c \to 0^+} \left[\frac{3}{2} x^{2/3} \right]_c^1$$

$$= \lim_{c \to 0^+} \left(\frac{3}{2} - \frac{3}{2} c^{2/3} \right)$$

$$= \frac{3}{2} - 0$$

Answer: $\quad = \dfrac{3}{2}$

Convergence or Divergence

Improper integrals **converge** if the *limit* = <u>finite number</u> (e.g., 1 or -30)

Improper integrals **diverge** if the *limit* = <u>infinite number</u> (e.g., ∞ or -∞)

Problem 13

Evaluate $\int_{-1}^{8} \dfrac{1}{3\sqrt{x}}\, dx$. The integrand is undefined at $x = 0$.

1. **Split integral into two at the undefined point:**

$$\int_{-1}^{8} \frac{1}{3\sqrt{x}} \, dx = \int_{-1}^{0} \frac{1}{3\sqrt{x}} \, dx + \int_{0}^{8} \frac{1}{3\sqrt{x}} \, dx$$

2. **Turn each integral into a *limit*:**

$$= \lim_{c \to 0^-} \int_{-1}^{c} \frac{1}{3\sqrt{x}} \, dx + \lim_{c \to 0^+} \int_{c}^{8} \frac{1}{3\sqrt{x}} \, dx$$

3. **Use reverse power rule, substitute, and evaluate:**

$$= \lim_{c \to 0^-} \left[\frac{3}{2} x^{2/3} \right]_{-1}^{c} + \lim_{c \to 0^+} \left[\frac{3}{2} x^{2/3} \right]_{c}^{8}$$

$$= \lim_{c \to 0^-} \left(\frac{3}{2} c^{2/3} - \frac{3}{2} \right) + \left(\lim_{c \to 0^+} 6 - \frac{3}{2} c^{2/3} \right)$$

$$= \quad -\frac{3}{2} \qquad\qquad + \qquad 6$$

Answer: = 4.5

Note: $(+)(-) = (-),$ so $+\frac{3}{2} - \frac{3}{2} = \frac{-3}{2}$ and $\frac{3}{2}(8) = \frac{24}{2} = \frac{12}{2} = 6$

So this improper integral *converges.*

Problem 14

Evaluate improper integral:

$$\int_{1}^{\infty} \frac{1}{x^2} \, dx, \text{ when c approaches infinity or negative infinity.}$$

1. **Turn integral into a** *limit*:

$$\int_1^\infty \frac{1}{x^2}\, dx \quad = \quad \lim_{c \to \infty} \int \frac{1}{x^2}\, dx$$

2. **Use reverse power rule, substitute, and evaluate:**

$$= \quad \lim_{c \to \infty} \left[\frac{-1}{x}\right]_1^c$$

$$= \quad \lim_{c \to \infty} \left(-\frac{1}{c} - \left(\frac{-1}{1}\right)\right)$$

$$= \quad 0 - (-1)$$

Answer: $= \quad\quad 1$

Note: (-) (-) = (+)

So this improper integral *converges.*

Problem 15

Evaluate improper integral:

$$\int_1^\infty \frac{1}{x}\, dx, \text{ when c approaches infinity or negative infinity.}$$

1. **Turn integral into a** *limit*:

$$\int_1^\infty \frac{1}{x}\, dx = \lim_{c \to \infty} \int_1^\infty \frac{1}{x}\, dx$$

2. Use reverse power rule, substitute, and evaluate:

$$= \lim_{c \to +\infty} [\ln x]_1^c$$

$$= \lim_{c \to +\infty} (\ln c - \ln 1)$$

$$= \infty - 0$$

Answer: $= \quad \infty$

Note: $\ln = \log_e$; thus, log base e (e ≈ 2.7182818 . . .)
So this improper integral *diverges*.

Problem 16

Evaluate: $\int_{-\infty}^{\infty} \dfrac{1}{x^2+1} \, dx$

1. Split integral into two:

$$\int_{-\infty}^{\infty} \frac{1}{x^2+1} \, dx = \int_{-\infty}^{0} \frac{1}{x^2+1} \, dx + \int_{0}^{\infty} \frac{1}{x^2+1} \, dx$$

2. Turn each integral into a *limit* and derive:

$$= \lim_{x \to -\infty} \int_{c}^{0} \frac{1}{x^2+1} \, dx + \lim_{x \to \infty} \int_{0}^{c} \frac{1}{x^2+1} \, dx \, *$$

3. **Evaluate, substitute, and add up results:**

$$= \lim_{c \to +\infty}\left[\arctan(x)\right]_c^0 + \lim_{c \to +\infty}\left[\arctan(x)\right]_0^c$$

$$= \lim\left(\arctan(0) - \arctan(c)\right) + \lim\left(\arctan(c) - \arctan(0)\right)$$

$$= \left(0 - \left(\frac{-\pi}{2}\right) + \pi - 0 \right)$$

Answer: = π

* Derivative of $\dfrac{1}{x^2 + 1}$ is $\dfrac{d}{dx}$ arctanx $= \dfrac{1}{x^2 + 1}$

If either one of the "half" integral *diverges*, the whole integral *diverges*.

Comparison Tests

There are three comparison tests:

Direct Comparison Test
Limit Comparison Test
Integral Comparison Test

Note: These three comparison tests can be seen in detail in appendix 3.

Infinity

Next to infinity, any finite thing amounts to *nothing*! Infinite series can go forever, as in the case of $0.1 + 0.01 + 0.001 + 0.0001 + \ldots$

Sequences and Series

Sequence is a <u>list</u> of numbers (terms), and an ***infinite sequence*** is an infinite list of numbers (terms).

List of numbers (terms)

$$\textit{Sequence} \quad = \quad \frac{1}{2}, \frac{1}{4}, \frac{1}{8}, \frac{1}{16}, \ldots$$

Infinite list of numbers (terms)

$$\textit{\textbf{Infinite Sequence}} \quad = \quad a_1, a_2, a_3, a_4, \ldots a_n{}^{*}$$

*   Term a_n is referred as "a sub n," and $\mathbf{a_n = \dfrac{1}{2^n}}$, and

$$\textbf{Infinite Sequence} \left\{ \frac{1}{2^n} \right\} = \frac{1}{2}, \frac{1}{4}, \frac{1}{8}, \frac{1}{16}, \ldots, \frac{1}{2^n}$$

a_n means "in the limit", and the terms of the *sequence* in the limit get smaller and smaller; and if it goes out far enough, the term can get *close* to zero. As n approaches *infinity* **but never gets there,** an gets *closer* and *closer* to zero.

Series is an <u>addition</u> of the infinite number of terms of a *sequence*.

$$Series \; = \; \frac{1}{2}+\frac{1}{4}+\frac{1}{8}+\frac{1}{16}+\ldots$$

Note: These sequences and series can be seen in detail in appendix 4.

Chapter 5

MYSTICISM AND TRANSCENDENTAL KNOWLEDGE

G reeks envisioned a vast world that is governed by invisible, nonmaterial principles transcending time and space. Then Pythagoreans thought of them as numbers and numerical relationships, and later they were thought as timeless ideas or forms. These assumptions are the foundation of modern science.

For millenniums people have been searching for truth by studying the complexity of the universe. Looking at a myriad of different forms and rhythms, they have seen patterns, relationships, and signs. But still there is an elusive clue or formula that humans have not been able to define that is a key to the unifying principle.

Using the language of comparison and mathematical relationships, there is a way to apply what is called the Golden Proportion (Phi $= (1 + \sqrt{5/2}) \approx 1.61803 \ldots$) to life's intrinsic mysteries by placing the larger next to the smaller and holding them both up to the *whole*. By doing that, a new dimension unfolds with an elusive mysterious code of perfect relationship of balance, harmony, and symmetry.

The Golden Proportion is easily described as *the whole is to the larger in exactly the same proportion as the larger is to the smaller.* It simply means that there is a relationship that can be proven with numbers, and that its principles of harmony are acknowledged as fundamental truths in the cosmic world and in the proportions of our own bodies.

In order to create a proportion, we place an equal sign between *two ratios*; and in order to describe a *ratio* with numbers, we use a line to divide its two parts. For example,

$$\text{Proportion: } \o = \left(\frac{1+\sqrt{5}}{2} \right) \quad \text{Ratio} = \frac{8}{2}$$

Golden Proportion or Golden Mean |_____|_____|
                                                    A   C   B

The Golden Mean portrays the relationship that the *whole line* is to the *larger segment* as the *larger segment* is to the *smaller segment.* In other words, the ratio of the whole line AB to its larger part AC is the same as the ratio of the larger part AC to the smaller part CB. When calculated mathematically, this gives a ratio of approximately 1.61803 to 1, which can also be expressed as ($\frac{1 + \sqrt{5}}{2}$) or ø.

History is a living proof of the long list of philosophers, architects, mystics, scientists, mathematicians, poets, artists, etc., who have experienced an admiration for the harmony of this Golden Proportion and who have taken its secrets to discover ever new and more exciting properties. Used in architecture, it has produced buildings of great symmetry. We can see the Golden Proportion being used in the Parthenon in Athens, in the Great Pyramids of Egypt, and in European gothic cathedrals.

Looking at the relationship of macrocosm and microcosm, the Golden Proportion defines the larger and the smaller in their most intimate relationship and links them in such a way that produces a mirror effect enabling us to see the large in the small and vice versa. They are not separate; they are related. It makes us catch a glimpse of the infinite process and gives us a hint of the universal law.

The circle is envisioned as the parent of all subsequent shapes. The Greeks used the term *monad* to describe the principles represented by the circle, derived from the root *monas,* which means *oneness.* Ancient mathematical philosophers referred to the nomad as the mother, the first, the seed, the essence, the unity, the builder, and the foundation.

The center of the circle is the source. In antiquity, it was regarded as the source and center of all things, and it was perceived as beyond understanding. It was said to be the unknowable; but like the seed, the ovum—the egg produced by the ovary of a plant, a woman, or female animal—became the source of anything from which life emerged. The center would expand and would fulfill itself as a circle.

Mathematical philosophers in ancient times believed that nothing exists without a center around which it revolves. They noticed that no matter how many times one is multiplied by itself, the result is always one. How does one become the many? The answer is because it is a reflection. The circle replicates itself by contemplating itself. This process is developed further in geometry as the birth of the lines that connects their centers. The Greek term for the principles represented by the circle was *monad* from the root *monas* or oneness.

The principle of duality was called *dyad* by the Greek philosophers, and it was also referred as an "illusion" because it implies a separation from the wholeness and a yearning to return to oneness. It occurs everywhere and is at the root of our notion of being apart from each other, from nature, and from our own divine source.

In the metaphor of arithmetic, the dyad reveals itself as the one and the many between the *monad* and all other numbers. This metaphor was seen as a symbolic pathway to the journey of self-discovery. It has also been associated with fertility and the Divine Feminine, and it relates to the spiritual journey to the passage of birth.

It was Plato who described a reality in which being and existence are bound together as one in which reality is inseparable from the natural harmony; and proportion that exist throughout creation as well as the relationship of all parts within the whole and to the whole is reflected in the relationship between us and the universe.

One can see in nature the immutable laws that play out in the unfolding of life with their remarkable simplicity and in their complex subtleties. When we feel the wind, the flow of water, the changing of color of dawn and dusk, we can see Mother Nature's laws, and we can experience the connectedness with her.

The spiral is an essential and one of the most significant tools of Mother Nature. We can see it in the sky, the water, the wind, in a lettuce or cabbage, or the seeds of a daisy. Spirals are all around us; some are more obvious than others. From galaxies to embryos, the spiral represents one of Mother Nature's more dramatically proportionate messages in that it arises out of the harmonization of opposites. It is the middle way, the path of least resistance, always finding the perfect balance. It unwinds around a point while moving ever farther from that point.

The spiral embodies the dynamic principles of regeneration whose imprint in life is of symmetrical and balanced growth. Life is either expanding, growing, being drawn out, or it is dissolving, diminishing, collapsing. These unfolding patterns we can see in roses, in plants, in our own bodies, and even in the calm eye of a storm, which is the center of gravity around which wind and water are expanding and contracting and the whole storm is balanced.

The eye of every spiral is a dynamic place where all opposites meet and where life and death are one. All forces that create growth and keep it in balance are at work in the center—the source. Throughout nature, from the incredibly vast to the infinitesimally small, we can see things moving with the different principles of balance along with an inborn tendency to move toward an equilibrium.

The Golden Proportion is in itself an expression of our intimate relationship with wholeness. The problematic paradox of western tradition is a belief in the so-called God and the division in the world we experience, which separate us from the whole. The experience of unity in existence is fundamental for humans since the division of oneness is impossible. However, in our modern world, it is impossible to perceive it through religious antagonisms and divisions.

Appendix 1

THE LIMIT OF RIEMANN SUMS

The Limit of Riemann Sums

5 Approximating Areas with 5 Rules

1. Left Rectangles Sum Rule   — Riemann Sum
2. Right Rectangles Sum Rule   — Riemann Sum
3. Midpoint Rectangles Sum Rule   — Riemann Sum
4. Trapezoid Rule
5. Simpson's Rule

* All three sums: *left*, *right*, and *midpoint* are called *Riemann sums* after the German mathematician G. F. B. Riemann (1826-66).

1. Sum of Left Rectangles

Sum of Left Rectangles **Formula**

(L = left and n = number of rectangles)

Width Heights

$$Ln = \frac{b-a}{n} \left[f(x0) + f(x1) + f(x2) + \ldots + f(xn-1) \right]$$

Note: The **function values** are the **heights** of the **rectangles.**

Left Rectangle **Rule**

An approximation of the exact area under a curve between a and b,

$$\int_a^b f(x) \; dx$$

can be obtained with a sum of *left rectangles* given by the *formula.* In general, the more rectangles, the better the estimate.

Graph example: $f(x) = x^2 + 1$

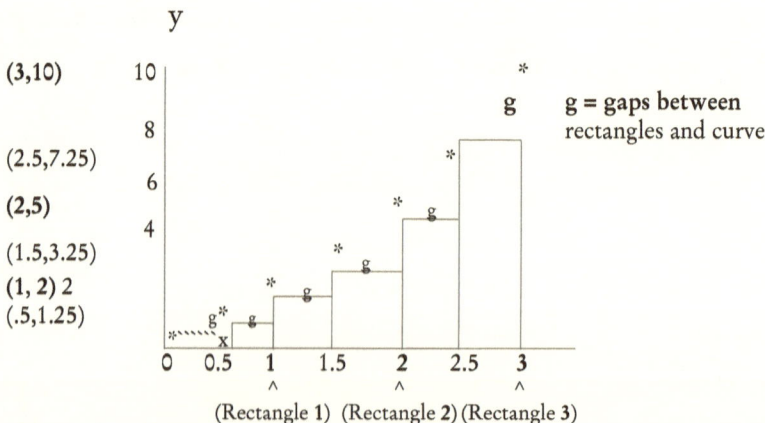

(Rectangle 1)   (Rectangle 2) (Rectangle 3)

Note: If there are just **3** rectangles and the upper left corner of each rectangle touches the curve, they represent a *left sum.* In this case, each rectangle has a width of 1 and the **height** of each is given at the height of the function at the rectangle's left edge. Adding the three areas gives a total of **8,** which is an underestimate of the total area under the curve because of the three gaps between the rectangles and the curve. For a better estimate, it is necessary to double the number of rectangles to **6** as it is shown in the above graphic.

	Height:	Area: (height x width)
Rectangle **I**	$f(0) = 0^2 + 1 = 1$	1 x 1 = 1
Rectangle **2**	$f(1) = 1^2 + 1 = 2$	2 x 1 = 2
Rectangle **3**	$f(2) = 2^2 + 1 = 5$	5 x 1 = 5
	Total	8

Table showing estimates under

$$f(x) = x^2 + 1$$

(given by increasing numbers of "left" rectangles)

Number of Rectangles:	Approximate Area:
3	8
6	9.875
12	~ 10.906
24	~ 11.445
48	~ 11.721
96	~ 11.860
192	~ 11.930
384	~ 11.965

Six Left Rectangles

The width of each rectangle equals the length of the total span from 0 to 3, that is 3-0 or **3** divided by the number of rectangles, yielding **6.** That is exactly what the <u>b—a</u> does in the formula. Therefore, $3-0 = \dfrac{3}{6} = \dfrac{1}{2} \dfrac{(3 \times 1 = 3)}{(3 \times 2 = 3)}$

2. Sum of Right Rectangles

Sum of Right Rectangles Formula
(**L** = left and **n** = number of rectangles)

Width	Heights
$Rn = \dfrac{b-a}{n}$	$[f(x1) + f(x2) + f(x3) + \dots + f(xn)]$

Note: The **function values** are the **heights** of the **rectangles.**

*Right Triangle **Rule***

An approximation of the exact area under a curve between **a** and **b,**

$$\int_a^b f(x) \; dx$$

can be obtained with a sum of *right rectangles* given by the *formula.* In general, the more rectangles, the better the estimate.

Graph example: $f(x) = x^2 + 1$

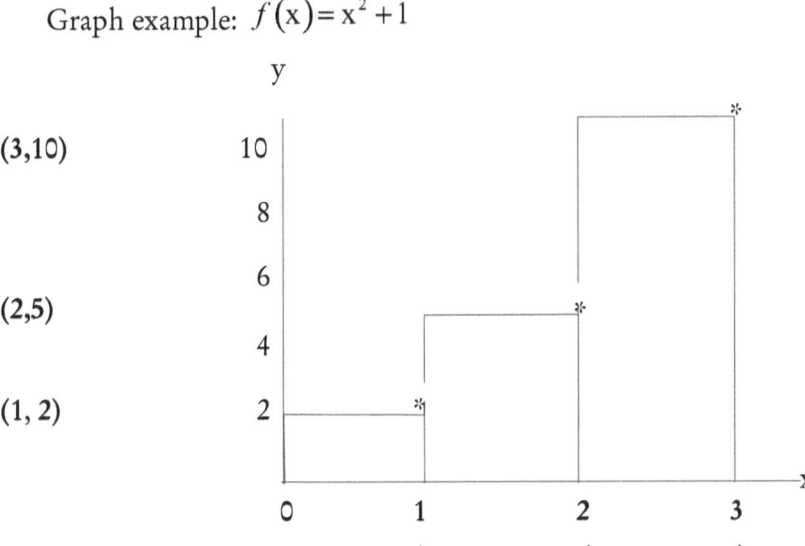

(Rectangle 1)   (Rectangle 2)(Rectangle 3)

Table showing estimates under

$$f(x) = x^2 + 1$$

(given by increasing numbers of "right" rectangles)

Number of Rectangles:		Approximate Area:
3		17
6		14.375
12	~	13.156
24	~	12.570
48	~	12.283
96	~	12.141
192	~	12.070
384	~	12.035

If we compare the areas and total of the *left rectangle sum* and the *right rectangles sum,* we can get a better picture.

Three *left rectangles:* $1 + 2 + 5 = 8$

Three *right rectangles:* $2 + 5 + 10 = 17$

The sums of the areas are the same *except* for the *leftmost rectangle* and the *right most rectangle.* The difference between them is 9 (17 minus 8 = 9).

3.   Sum of Midpoint Rectangles

The third method to estimate areas with rectangles is to make each rectangle cross the curve at the *midpoint* of its topside. A midpoint sum is a much better approximated area than either a *left* or *right* sum.

Sum of Midpoint Rectangles Formula
(L = left and n = number of rectangles)

Width   **Heights**

$$\mathbf{Mn} = \frac{b - a}{n}\left[f\left(\frac{x0 + x1}{2}\right) + f\left(\frac{x1 + x2}{2}\right) + f\left(\frac{x2 + x3}{2}\right) + \ldots + f\left(\frac{xn - 1 + xn}{2}\right)\right]$$

Note: The **function values** are the **heights** of the **rectangles.**

Midpoint **Rule**

An approximation of the exact area under a curve between **a** and **b**,

$$\int_a^b f(x)\ dx$$

can be obtained with a sum of *midpoint rectangles* given by the *formula.* In general, the more rectangles, the better the estimate.

Graph example: $f(x) = x^2 + 1$

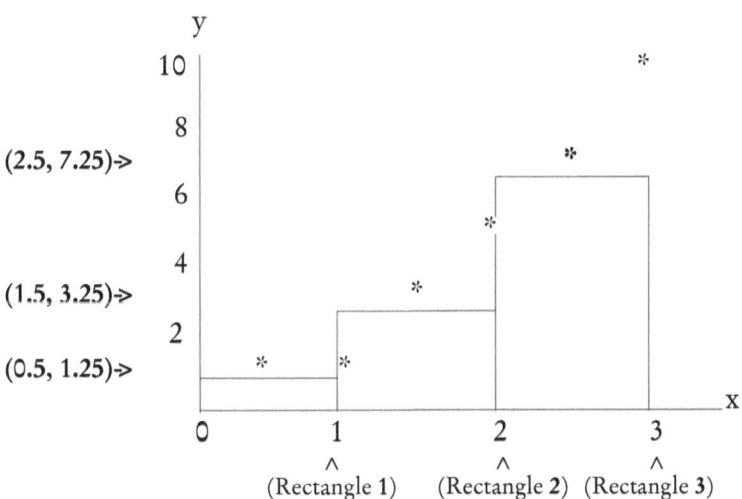

(2.5, 7.25)->

(1.5, 3.25)->

(0.5, 1.25)->

(Rectangle 1)   (Rectangle 2) (Rectangle 3)

Table Showing Estimates Under

$$f(x) = x^2 + 1$$

(given by increasing numbers of "midpoint" rectangles)

Number of Rectangles:		Approximate Area:
3		11.75
6		11.9375
12	~	11.9845
24	~	11.9961
48	~	11.9990
96	~	11.9998
192	~	11.9999
384	~	11.99998

Note: The *left*, *right*, and *midpoint* sums are all heading toward **12**; and if it was possible to slice up the area into an *infinite* number of rectangles, it will give the *exact area of* **12**.

4. Sum of Trapezoid

Besides using left, right and midpoint rectangles, a fourth method can be used by approximating area with trapezoids and parabola-topped "trapezoids" instead of rectangles.

Sum of Trapezoids Formula
(**L** = left and **n** = number of trapezoids)

Width   Heights

$$\mathbf{Tn} = \frac{\mathbf{b - a}}{\mathbf{2n}} \left[f(x0) + 2f(x1) + 2f(x2) + \ldots + 2f\left(xn - 1 + f(xn)\right) \right]$$

Note: x0 = a and x1 through xn are the equally-spaced x coordinates of the right edges of trapezoids 1 through n.

Trapezoid *Rule*

An approximation of the exact area under a curve between **a** and **b**,

$$\int_{a}^{b} f(x) \; dx$$

can be obtained with a sum of *trapezoids* given by the *formula*. In general, the more trapezoids, the better the estimate.

Graph example: $f(x) = x^2 + 1$. Three trapezoids approximate the area.

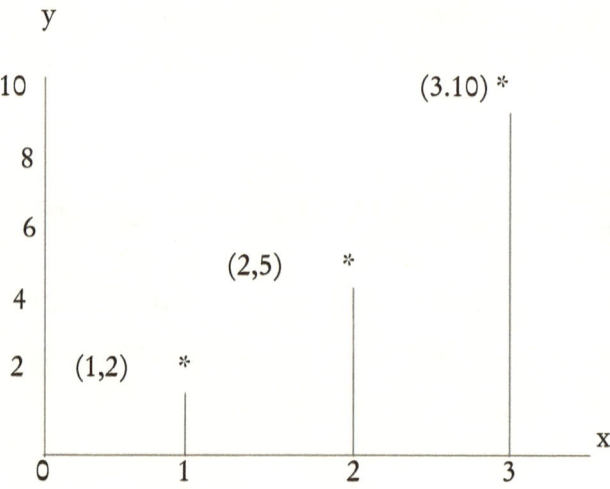

Table Showing Estimates Under

$$f(x) = x^2 + 1$$

(between 0 and 3, given by increasing numbers of trapezoids)

Number of Trapezoids:		Approximate Area:
3		12.5
6		12.125
12	~	12.031
24	~	12.008
48	~	12.002
96	~	12.0004
192	~	12.0001
384	~	12.00003

Note: The *midpoint* table of estimates lists an area estimate of 11.9990 for 48 midpoint rectangles. The area estimate with 48 trapezoids given above is 12.002.

5. Simpson's Sum of Parabola-topped "Trapezoids"

The difference between trapezoids and parabola-topped "trapezoids" is that the latter have curved, parabolic tops instead of having slanting tops.

Simpson's rule is by far the *most accurate approximation method.* Indeed it gives the exact area for any polynomial function with degrees of three or less. It gives a much better estimate than the midpoint of the trapezoid rule.

Sum of Parabola-topped "trapezoids" Formula
(S = Simpson, and **n** = twice the number of trapezoids)

Width Heights

$$S_n = \frac{b-a}{3n}\left[f(x0) + 4f(x1) + 2f(x2) + 4f(x3) + 2f(x4) + \ldots + 4f(xn-1) + f(xn)\right]$$

Note: x0 through xn are the n + 1 evenly spaced points from a to b.

Simpson's **Rule**

An approximation of the exact area under a curve between **a** and **b**,

$$\int_a^b f(x)\ dx$$

can be obtained with a sum of *parabola-topped "trapezoids"* given by the **formula.** In general, the more "trapezoids," the better the estimate.

Simpson's sum is like an average of a midpoint sum and a trapezoid sum. The only difference is that we use the midpoint sum twice in the average. We can obtain the Simpson's rule using the following average if we already have the midpoint and the trapezoid sums for some number of rectangles or trapezoids.

$$S2n \;=\; \frac{Mn \;+\; Mn \;+\; Tn}{3}$$

Notice that the above formula always have the same number of rectangles, trapezoids, and Simpson's rule "trapezoids." If we do not have the midpoint and trapezoid sums in order to use the above average, we have to use Simpson's rule.

Writing Riemann Sums with Sigma Notation

Right Sum Formula: **Width**   **Heights**

$$Rn \;=\; \frac{b-a}{n} \; \left[f(x1) + f(x2) + f(x3) + \ldots + f(xn) \right]$$

With Sigma Notation:

$$Rn \;=\; \sum_{i=1}^{n} \left[f(xi) \cdot \left(\frac{b-a}{n} \right) \right]$$

Finding the *exact area with the Definite Integral*

Integral symbol gives *total area* under a curve between a and b.

$$\int_a^b f(x) \ dx$$

Definite Integral gives the *exact area* under a curve between a and b.

$$\int_a^b f(x) \ dx \;=\; \lim_{n \to \infty} \sum_{i=1}^{n} \left[f(xi) \cdot \frac{(b-a)}{n} \right]$$

The summation above is identical to the formula for **n right rectangles,** Rn, given before. The only difference is that we take the limit of that formula as the number of rectangles approaches *infinity* (∞).

Here we finally have the *exact area* under $x^2 + 1$ between a and b.

$$\int_0^3 (x \ + \ 1) \ dx \;=\; \lim_{n \to \infty} \sum_{i=1}^{n} \left[f(xi) \cdot \frac{(b-a)}{n} \right]$$

$$= \; \lim_{n \to \infty} \left(12 + \frac{27}{2n} + \frac{9}{2n^2} \right)^{*}$$

$$= \; 12 + \frac{27}{2 \cdot \infty} + \frac{9}{2 \cdot \infty^2}$$

$$= \; 12 + \frac{27}{\infty} + \frac{9}{\infty}$$

$$= \; 12 + 0 + 0^{*}$$

$$= \; 12$$

*   Remember that in a limit problem, any number *divided* by *infinity* equals zero.

Here is the *left rectangle* limit:

$$\int_0^3 \left(x^2 + 1\right)\, dx = L\infty = \lim_{n \to \infty} \left(12 - \frac{27}{2n} + \frac{9}{2n^2}\right)$$

$$= \quad 12 - \frac{27}{2 \cdot \infty} + \frac{9}{2 \cdot \infty^2}$$

$$= \quad 12 - \frac{27}{\infty} + 9$$

$$= \quad 12 + 0 + 0$$

$$= \quad 12$$

Here is the *midpoint rectangle* limit:

$$\int_0^3 (x + 1)\, dx \ = \ M\infty \ = \ \lim_{n \to \infty} \left(12 - \frac{9}{4n^2}\right)$$

$$= \quad 12 - \frac{9}{4 \cdot \infty^2}$$

$$= \quad 12 - \frac{9}{\infty}$$

$$= \quad 12 - 0$$

$$= \quad 12$$

Appendix 2

SIX TECHNIQUES TO FIND ANTIDERIVATIVES

Six Techniques to Find Antiderivatives

First Technique

Reverse Rules for Antiderivative Method (*)

Second Technique

Trying and Checking Method

Third Technique

Substitution Method

Fourth Technique

Integration by Parts Method

Fifth Technique

Trigonometric Integrals Method

Sixth Technique

Trigonometric Substitution Method

Seven Technique

Partial Fractions Method

First Technique (*)
Reverse Rules for Antiderivatives Method

This first technique has been given in detail previously (see page 133)

Second Technique
Trying and Checking Method

This method works when the *integrand* (expression placed after the *integral* symbol not including the *dx)* is easy enough to anti differentiate. For example,

Find the antiderivative of cos(2x):

Derivative: Anti-derivative:

$$\frac{d}{dx}\ \cos(2x)\,dx \qquad\qquad \int \cos(2x)\,dx\ =\ \frac{1}{2}\sin(2x)+C^{*}$$

* *Multiplicative Inverse* or *Reciprocal* of *2* or $\dfrac{2}{1}\ =\ \dfrac{1}{2}$

This *reciprocal of 2* comes from the coefficient of x-the **2** in **(2x)**.

Third Technique
Substitution Method

The substitution method works when the **integrand** contains a function and the *derivative of the function's* **argument.**

Note: An *argument* is what is inside of a *function,* as in a function like $\sqrt{4x}$, the **4x** if called the *argument.*

Find the anti derivative of: $\int 2x \cos(x^2)dx$

Argument of $\cos(x^2)$ is: (x^2)

Original Function: Derivative: Antiderivative:

$\sin(x)^2$ $\dfrac{d}{dx} 2x \cos(x)^2 dx$ $\int 2x\cos(x)^2 dx = \sin(x)+C$

1. Set **u**: $\int 2x \cos(x^2)dx$

$$\mathbf{u} \quad = \quad (\mathbf{x^2})$$

$$\frac{\mathbf{du}}{\mathbf{2x}} \quad = \quad \mathbf{dx}$$

2. Cancel **2x** and substitute: $\int \cos \ \mathbf{u} \ \mathbf{du} \ = \ \sin \ \mathbf{u}+C$

   Answer: $\int 2x\cos(x^2)dx = \sin(x^2)+C$

Fourth Technique
Integration by Parts Method

Integration by parts simply means to transform a *difficult integral* into a simple product **minus** an *easy integral.*

Formula

$$\int \mathbf{udv} \ = \ \mathbf{uv} \ -\int \mathbf{vdu}$$

Note: An easy way to remember the formula is to notice that **uv** are in alphabetical order, except at the end where the **uv** are reversed.

$$\int \quad \text{udv} \qquad = \quad \text{uv} \quad -\int \quad \text{vdu}$$

$$\int \sqrt{x}\ln(x)\ dx \ = \ \ln(x)\cdot\frac{2}{3}x^{3/2} \ -\int \frac{2}{3}x^{3/2}\cdot\frac{1}{x}\ dx$$

Let **u** *be* $\qquad \ln(x) \qquad\qquad$ *Let* **v** *be* $\frac{2}{3}x^{3/2}$ *

Let $\int$ **dv** *be* $\int \sqrt{x}\ dx$ *Let* **du** *be* $\frac{1}{x}\ dx$

*   Remember ln = $\log_e$ (e ≈ 2.72) and *reverse power rule.*

Example:

$$\int \quad \text{udv} \qquad = \quad \text{uv} \quad -\int \quad \text{vdu}$$

$$\int \sqrt{x}\ln(x)\ dx \ = \ \ln(x)\cdot\frac{2}{3}x^{3/2} \ -\int \frac{2}{3}x^{3/2}\cdot\frac{1}{x}\ dx$$

$$= \ \frac{2}{3}x^{3/2}\ \ln(x) \ -\frac{2}{3}\int x^{3/2}\ dx$$

$$= \ \frac{2}{3}x^{3/2}\ \ln(x) \ -\frac{2}{3}\left(\frac{2}{3}x^{3/2}+C\right)$$

$$= \ \frac{2}{3}x^{3/2}\ \ln(x) \ -\frac{4}{9}\int x^{3/2}-\frac{2}{3}C^{*}$$

$$= \ \frac{2}{3}x^{3/2}\ \ln(x) \ -\frac{4}{9}\int x^{3/2}+C$$

*   Replace $-\frac{2}{3}C$ for +C, because $-\frac{2}{3}$ times any number is just any number.

Fifth Technique
Trigonometric Integrals Method

In order to use this technique, we need an integrand that contains just one of the six trig functions, namely, sine, cosine, tangent, cosecant, secant, and cotangent. If the integrand contains two trig functions, the two should be one of the following pairs: sine and cosine, secant with tangent, or cosecant with cotangent. If the trig functions are different from the above mentioned, we can convert them by using trig identities:

Reciprocal Identities: Quotient Identities: Pythagorean Identities:

$$\csc \theta = \frac{1}{\sin \theta} \qquad \tan \theta = \frac{\sin \theta}{\cos \theta} \qquad \sin^2 \theta + \cos^2 \theta = 1$$

$$\sec \theta = \frac{1}{\cos \theta} \qquad \cot \theta = \frac{\cos \theta}{\sin \theta} \qquad \tan^2 \theta + 1 = \sec^2 \theta$$

$$\cot \theta = \frac{1}{\tan \theta} \qquad\qquad\qquad\qquad 1 + \cot^2 \theta = \csc^2 \theta$$

After doing the necessary conversions, we have one of the following three cases:

$$\int \sin^m (x) \cos^n (x) \, dx$$

$$\int \sec^m (x) \tan^m (x) \, dx$$

$$\int \csc^m (x) \cot^m (x) \, dx$$

Where either m or n is a *positive integer.* It is generally preferable to have positive powers of trig functions rather than negative powers; therefore, it is convenient to convert:

$$\int \sin^{-2}(x) \tan^{-2}(x) \, dx \quad \text{into:}$$

$$\int \csc^2(x) \cot^2(x) \, dx$$

Pythagorean identity says that for *any angle x:*

$$\sin^2 (x) + \cos^2 (x) = 1$$

Therefore,

$$\sin^2(x) \qquad\qquad = 1 - \cos^2(x) \qquad \text{and}$$

$$\cos^2(x) = 1 - \sin(x).$$

Example:

Integrate: $\quad \int \sin^3(x) \cos^4(x)\, dx$ **(sine power is *odd* and *positive*)**

Lop off : $\quad \int \sin^3(x) \cos^4(x)\, dx = \int \sin^2(x) \cos^4(x) \sin(x)\, dx$

Convert sines

to cosines: $\quad \int \sin^2(x) \cos^4(x) \sin(x)\, dx$

$$= \int (1 - \cos^2(x)) \cos^4(x) \sin(x)\, dx$$

$$= \int (\cos^4(x) - \cos^4(x)) \sin(x)\, dx$$

3. **Twitch integral and**
   substitute: $\quad = -\int (\cos^4(x) - \cos^6(x)) (-\sin(x) dx)$

$$
\begin{aligned}
\text{Let}\quad u &= \cos(x)\\
du &= -\sin(x)\, dx\\
\frac{du}{dx} &= -\sin(x)\\
&= -\int (u^4 - u^6)\, du
\end{aligned}
$$

4. **Use reverse power rule:** $\quad = -\frac{1}{5} u^5 + \frac{1}{7} u^7 + C$

Answer: $\quad = \frac{1}{5} \cos^7(x) - \frac{1}{5} \cos^5(x) + C$

Sixth Technique
Trigonometric Substitution Method

Radical Table

$$\sin(\theta) \quad \leftrightarrow \quad \sqrt{a^2 - u^2}$$

$$\sec(\theta) \quad \leftrightarrow \quad \sqrt{u^2 - a^2}$$

$$\tan(\theta) \quad \leftrightarrow \quad \sqrt{u^2 + a^2}$$

Example: **Tangent**

1.

Integrate: $\displaystyle\int \frac{dx}{\sqrt{9x + 4^2}}$ Rewrite: $\displaystyle\int \frac{dx}{\sqrt{(3x)^2 + 2^2}}$

$$= \sqrt{u^2 + a^2} \quad \tan(\theta) = \frac{O}{A}$$

$$= O + A$$

$$\tan(\theta) = \frac{3x}{2} \text{ for } x$$

Key: $\dfrac{u}{a}$

2. **Draw** right triangle and solve for **x**:

$$\left(\tan(\theta) = \frac{u}{a}, \text{ which is} \tan(\theta) \frac{O}{A} = \frac{3x}{2} \right)$$

Therefore, $\tan(\theta) = \dfrac{3x}{2}$, so $3x = 2\tan(\theta)$

$\sqrt{9x+4}$ or $\sqrt{(3x+2)}=$

H (hypotenuse)

3x = O (Opposite)

2 = A (Adjacent)

Key: $\dfrac{\sqrt{}}{a}$

3. **Differentiate** and solve for **dx**:

$$\left(sec\,(\theta) = \frac{u}{a}, \text{ which is } sec\,(\theta)\frac{H}{A} = \frac{\sqrt{9}+4}{2} \right)^2$$

Therefore, $sec\,(\theta) = \dfrac{\sqrt{9x}^2+4}{2}$, so $\sqrt{9x}^2+4 = 2sec\,(\theta)$

$$\frac{3x}{2} \quad = \quad \tan\,(\theta)$$

$$3x \quad = \quad 2\tan(\theta)$$

$$x \quad = \quad \frac{2}{3}\tan(\theta)$$

$$\frac{dx}{d\theta} \quad = \quad \frac{2}{3}sec^2\,(\theta)$$

$$dx \quad = \quad \frac{2}{3}sec^2(\theta)\,d\theta$$

4. **Integrate:** $\displaystyle\int \dfrac{dx}{\sqrt{9x}^2+4} \quad = \quad \displaystyle\int \dfrac{\frac{2}{3}sec^2\,(\theta)\,d\theta}{2\,sec(\theta)}$

$$= \quad \frac{1}{3}\int\ sec(\theta)\,d\theta^*$$

$$= \quad \frac{1}{3}ln\big|\,sec(\theta)+\tan(\theta)\,\big| \ +C$$

*   Integral $\displaystyle\int sec\,x\,(dx) = ln|sec x + \tan x| + C$

5. Substitute:

$$= \frac{1}{3}\ln\left|\sqrt{9x+4} + 3x\right| - \frac{1}{3}\ln 2 + C*$$

* Eliminate constant $-\frac{1}{3}\ln 2$

Answer:

$$= \frac{1}{3}\ln\left|\sqrt{9x+4} + 3x\right) + C$$

Example: **Sine**

1.

Integrate: $\displaystyle\int \frac{dx}{x^2\sqrt{16-x^2}}$ Rewrite: $\displaystyle\int \frac{dx}{x^2\sqrt{4^2-x^2}}$

$$= \sqrt{a^2 - u^2} \quad \sin(\theta) = \frac{O}{H}$$

$$= \quad H - O$$

$$\sin(\theta) = \frac{x}{4} \quad \text{for} \quad x$$

Key: $\dfrac{u}{a}$

2. Draw right triangle and solve for **x:**

$$\left(\sin(\theta) = \frac{u}{a}, \text{ which is } \sin(\theta)\frac{O}{H} = \frac{x}{4} \right)$$

Therefore, $\sin(\theta) = \dfrac{x}{4}$, so $x = 4\sin(\theta)$

$4 = H$ (Hypotenuse)

$x = O$ (Opposite)

$)\theta$

$\sqrt{16-x^2}\ =\ A$ (Adjacent)

Key: $\dfrac{\sqrt{\ }}{a}$

3. **Differentiate** and solve for **dx:**

$$\left(\sin(\theta) = \frac{u}{a},\ \text{which is}\ \cos(\theta)\frac{A}{H} = \frac{\sqrt{16-x^2}}{4}\right)$$

Therefore, $\cos(\theta) = \dfrac{\sqrt{16-x^2}}{4}$, so $\sqrt{16-x^2} = 4\cos(\theta)$

$$\frac{x}{4} = \sin(\theta)$$

$$x = 4\sin(\theta)$$

$$\frac{dx}{d\theta} = 4\cos(\theta)$$

$$dx = 4\cos(\theta)d\theta$$

4. **Integrate:** $\dfrac{dx}{x^2\ \sqrt{16-x^2}} = \displaystyle\int \dfrac{4\cos(\theta)\,d\theta}{\left(4\sin(\theta)^2\right)\ 4\cos(\theta)}$

$$= \int \frac{d\theta}{16\sin^2(\theta)}$$

$$= \frac{1}{16}\int \csc^2(\theta)\,d(\theta)$$

$$= -\frac{1}{16}\cot(\theta) + C^*$$

* Triangle shows that $\cot(\theta) = \dfrac{\sqrt{16 - x^2}}{x}$

5. **Substitute:** $= \dfrac{-1}{16} \cdot \dfrac{\sqrt{16 - x}}{x} + C$ *

Answer: $= -\dfrac{\sqrt{16 - x}}{16x} + C$

Seventh Technique
Partial Fractions Method

When using the partial fraction method, check that *integrand* is a **proper fraction**, For example,

Proper fraction: $\displaystyle\int \dfrac{5}{x^2 + x - 4}$ *

Improper fraction: $\displaystyle\int \dfrac{5}{x^2 + x - 4}$ *

* Proper fraction is when the *degree* of the <u>numerator</u> is *less* than the *degree* of the denominator.

Convert an *improper fraction* into a *proper fraction* with a simple division:

Division: $\dfrac{21}{4} = \begin{array}{r} 5\ . \\ 4\overline{)21} \\ \underline{20} \\ 1 \end{array}$

Quotient is **5**   Remainder is **1**   Divisor is **4**

$$= 5\,\frac{1}{4}$$

Division: $\int \dfrac{2x^3 + x^2 - 6}{x^3 - 3x - 2}$ dx $= x^3 - 3x - 2 \overline{\smash{\big)}\ 2x^3 + x^2 + 0x - 6}$

$$\begin{array}{r} 2 \quad\quad\quad\quad\quad\quad \\ x^3 - 3x - 2 \overline{\smash{\big)}\ 2x^3 + x^2 + 0x - 6} \\ \underline{2x - 6x^3 - 4} \\ x^2 + 6x - 2 \end{array}$$

Quotient is 2; remainder is $x^2 + 6x - 2$; divisor is $x^3 - 3x - 2$.

$$= \int 2\,dx + \int \frac{x + 6x - 2}{x - 3x - 2}\ dx$$

Example:

Integrate: $\int \dfrac{5}{x^2 + x - 6}$

1. Factor: $\dfrac{5}{x^2 + x - 6} = \dfrac{5}{(x-2)(x+3)}$

2. Break up fraction:

$$\frac{5}{(x-2)(x+3)} = \frac{A}{(x-2)} + \frac{B}{(x+3)}$$

3. Cross multiply both sides
   by the denominator: $5 = A(x+3) + B(x-2)$

4. Take linear factors roots.
   Plug into x for unknowns:

If $x = 2$ If $x = -3$

$5 = A(2+3) + B(2-2)$ $5 = A(-3+3) + B(-3-2)$

$5 = \dfrac{5A}{5}$ $5 = \dfrac{-5B}{5}$

$A = 1$ $B = -1$

5. **Plug results in the equation and up original integral into partial fractions:**

$$\int \frac{5}{x^2 + x - 6} \, dx \;=\; \int \frac{1}{(x-2)} \, dx \;+\; \frac{-1}{(x+3)} \, dx$$

$$= \; \ln|x-2| \; - \ln|x+3| \; + \; C$$

$$= \; \ln \frac{|x-2|}{|x+3|} \qquad\qquad + \; C$$

Appendix 3

THREE COMPARISON TESTS AND RATIOS AND ROOTS TESTS

Direct Comparison Test

Direct Comparison Test: Let $0 \leq a_n \leq b_n$ for all n.

If $\sum_{n=1}^{\infty} b_n$ *converges*, then $\sum_{n=1}^{\infty} a_n$ *converges*

If $\sum_{n=1}^{\infty} a_n$ *diverges*, then $\sum_{n=1}^{\infty} b_n$ *diverges*

Example 1: **Assess** the *convergence or divergence* of :

$$\sum_{n=1}^{\infty} \frac{1}{5 + 3^n}$$

1. **Determine** the *benchmark series*:

$$\sum_{n=1}^{\infty} \frac{1}{5+3^n} \text{ resembles geometric series } \left(r = \frac{1}{3}\right) \quad \sum_{n=1}^{\infty} \frac{1}{3^n}$$

2. **Rewrite** in standard *geometric series* form:

$$\sum_{n=1}^{\infty} \frac{1}{3^n} \quad = \quad \sum_{n=1}^{\infty} \frac{1}{3} \frac{(1)^n}{3}$$

Answer: Because $0 < |r| < 1$ this series *converges*, and because $\frac{1^3}{5+3^n} < \frac{1}{3^n}$ for all values of n, $\sum_{n=1}^{\infty} \frac{1}{5+3^n}$ also *converges*.

Example 2: **Assess** the *convergence or divergence* of :

$$\sum_{n=1}^{\infty} \frac{lnn}{n}$$

1. **Determine** the *benchmark series*:

$$\sum_{n=1}^{\infty} \frac{lnn}{n} \text{ resembles harmonic p-series } \sum_{n=1}^{\infty} \frac{1}{n}$$

Answer: Because *harmonic p-series diverges*, this series *diverges*, and because $\frac{lnn}{n} > \frac{1}{n}$ for all values of $n \geq 3$, $\sum_{n=1}^{\infty} \frac{lnn}{n}$ also *diverges*.

Note: $n \geq 3$ is *only* considered. For any of the *convergence/divergence* tests, we can disregard any number of terms of the beginning of a series because the first, say 10 or 1,000 or 1,000,000, terms of a series always sum up to a finite number; and therefore never have any effect on whether the series *converges* or *diverges*.

Limit Comparison Test

Limit Comparison Test: For 2 series, $\sum a_n$ and $\sum b_n$, if $a_n > 0$, $b_n > 0$, and $\lim\limits_{n\to\infty} \dfrac{(a_n)}{b_n} = L$, then either both series *converge* or both *diverge*.

Note: L is a *finite* and *positive* number.

Example 1: **Assess** the *convergence or divergence* of :

$$\sum_{n=2}^{\infty} \frac{1}{n^2 - lnn}$$

1. **Determine** the *benchmark series*:

$$\sum_{n=2}^{\infty} \frac{1}{n^2 - lnn} \text{ resembles } convergent \text{ } p\text{-}series \sum_{n=2}^{\infty} \frac{1}{n^2}$$

2. **Take the limit** of the ratio of the nth terms of the 2 series:

$$\lim_{n\to\infty} \frac{\dfrac{1}{n^2 - lnn}}{\dfrac{1}{n^2}}$$

$$= \lim_{n\to\infty} \frac{n^2}{n^2 - lnn}$$

3. **Use L'hopital Rule:**

$$= \lim_{n\to\infty} \frac{2n}{2n - \dfrac{1}{n}}$$

4. **Use L'hopital Rule again:**

$$= \quad \lim_{n \to \infty} \frac{2}{2 + \frac{1}{n^2}}$$

$$= \quad \frac{2^n}{2 + \frac{1}{\infty}}$$

$$= \quad \frac{2}{2 + 0}$$

$$= \quad \mathbf{1}$$

Answer: $\displaystyle\sum_{n=2}^{\infty} \frac{1}{n^2 - \ln n}$ *converges.*

And because the *limit* is *finite* and *positive* and the *benchmark p-series converges*, this series also converges.

Example 2: **Assess** the *convergence or divergence* of :

$$\sum_{n=1}^{\infty} \frac{5n^2 - n + 1}{n^3 + 4n + 3}$$

1. **Determine** the *benchmark series*:

$$\sum_{n=1}^{\infty} \frac{5n^2 - n + 1}{n^3 + 4n + 3} \text{ resembles } harmonic\ series \sum_{n=1}^{\infty} \frac{1}{n}^{*}$$

2. **Take the limit** of the ratio of the nth terms of the 2 series:

$$= \quad \lim_{n \to \infty} \frac{\dfrac{5n^2 - n + 1}{n^3 + 4n + 3}}{\dfrac{1}{n}}$$

$$= \quad \lim_{n \to \infty} \frac{5n^3 - n^2 + n}{n^3 + 4n + 3}$$

3. **Divide** numerator and denominator by n^3.

$$= \lim_{n \to \infty} \frac{5 - \dfrac{1}{n} + \dfrac{1}{n^2}}{1 + \dfrac{4}{n^2} + \dfrac{3}{n^3}}$$

$$= \frac{5 - \dfrac{1}{\infty} + \dfrac{1}{\infty}}{1 + \dfrac{4}{\infty} + \dfrac{3}{\infty}}$$

$$= 5$$

* To determine the *benchmark series*, take the highest power of n in both numerator and denominator—ignoring coefficients and all other terms:

$$\frac{5n^2 - n + 1}{n^3 + 4n + 3} \to \frac{n^2}{n^3} = \frac{1}{n}$$

Answer:

$$\sum_{n=1}^{\infty} \frac{5n^2 - n + 1}{n^3 + 4n + 3} \ diverges$$

And because the *limit* is *finite* and *positive* and the *benchmark p-series diverges*, this series also diverges.

Integral Comparison Test

If the integral *converges*, the series *converges*; and if the integral *diverges*, the series *diverges*, and if $f(x)$ is *positive, continuous,* and *decreasing*, and if $a_n = f(n)$, then $\sum_{n=1}^{\infty} a_n$ and $\int_1^{\infty} f(x)\, dx$ either both *converge* or both *diverge*.

Example 1: **Assess** the *convergence or divergence* of :

$$\sum_{n=2}^{\infty} \frac{1}{n \ln n}$$

1. **Compute** the companion improper integral with the same *limits of integration* as the index numbers of the summation:

$$\int_{2}^{\infty} \frac{1}{x \ln x} \, dx$$

$$\lim_{b \to \infty} \int_{2}^{\infty} \frac{1}{x \ln x} \, dx$$

2. **Substitute:**

($u = \ln x$ and $du = \underline{1} \, dx$). When $x = 2$, $u = \ln 2$, and when $x = b$, $u = \ln b$

$$= \lim_{b \to \infty} \int_{\ln 2}^{\ln b} \frac{1}{u} \, du$$

$$= \lim_{b \to \infty} \left[\ln u \right]_{\ln 2}^{\ln b}$$

$$\lim_{b \to \infty} \left(\ln \left(\ln b \right)_{\ln 2} - \ln \left(\ln 2 \right) \right)$$

$$= \left(\ln \left(\ln \infty \right) - \ln \left(\ln 2 \right) \right)$$

$$= \infty - \ln \left(\ln 2 \right)$$

$$= \infty$$

Answer: $\sum_{n=2}^{\infty} \dfrac{1}{n \ln n}$ *diverges.*

Because the integral diverges, the series diverges also.

Ratios and Roots Tests

If the answer is less than 1, the series *converges*; if it is more than 1, the series *diverges.*

The ratio test is specially helpful with series involving *factorials* like n! or where n is in the *power*. For example,

Factorial: $4! = 4 \cdot 3 \cdot 2 \cdot 1 = 24$ factorial fraction: $\dfrac{4!}{3!} = \dfrac{4 \cdot 3 \cdot 2 \cdot 1}{3 \cdot 2 \cdot 1} = 4$

Power: 4^n

Example 1: **Assess** the *convergence or divergence* of :

$$\sum_{n=0}^{\infty} \frac{3^n}{n!}$$

Take the limit of the *ratio* of the (n + 1) term to the nth term:

$$\lim_{n \to \infty} \frac{\dfrac{3^{n+1}}{(n+1)!}}{\dfrac{3^n}{n!}}$$

$$= \lim_{n \to \infty} \frac{3 \cdot n!}{(n+1)! \cdot 3^n}$$

$$= \lim_{n \to \infty} \frac{3}{(n+1)}$$

$$= \frac{3}{\infty + 1}$$

$$= 0$$

Answer: Because this limit is less than 1, $\displaystyle\sum_{n=0}^{\infty} \frac{3^n}{n!}$ *converges.*

Example 2: **Assess** the *convergence or divergence* of :

$$\sum_{n=1}^{\infty} \frac{n^n}{n!}$$

Take the limit of the *ratio*:

$$\lim_{n \to \infty} \frac{\dfrac{(n+1)^{n+1}}{(n+1)!}}{\dfrac{n^n}{n!}}$$

$$= \lim_{n \to \infty} \frac{(n+1)^{n+1} \cdot n!}{(n+1)! \cdot n^n}$$

$$= \lim_{n \to \infty} \frac{(n+1)^{n+1}}{(n+1) \cdot n^n}$$

$$= \lim_{n \to \infty} \frac{(n+1)^n}{n^n}$$

$$= \lim_{n \to \infty} \frac{(n+1)^n}{n}$$

$$= \lim_{n \to \infty} \left(1 + \frac{1}{n}\right)^n$$

$$\approx e$$

$$\approx 2.7182818\ldots$$

Answer: Because this limit is greater than 1, $\sum_{n=1}^{\infty} \dfrac{n^n}{n!}$ *diverges.*

The root test looks at a *limit*; and like the ratio test, it is specially helpful if the series involves nth powers.

Example 1: **Assess** the *convergence or divergence* of :

$$\sum_{n=1}^{\infty} \frac{e^{2n}}{n^n}$$

Assess square root:

$$\lim_{n \to \infty} \sqrt[n]{\frac{e^{2n}}{n^n}}$$

$$= \lim_{n \to \infty} \frac{e}{n^{n/n}}$$

$$= \lim_{n \to \infty} \frac{e^2}{n}$$

$$= \frac{e}{\infty}$$

$$= 0$$

Answer: Because this limit is less than 1, $\displaystyle\sum_{n=1}^{\infty} \frac{e^{2n}}{n^n}$ *converges.*

Note: Expressions listed in increasing order from "smallest" to "biggest":

$$n^{10}, 10^2, n!, n^n$$

Appendix 4

SEQUENCES AND SERIES

Sequences and Series

A *sequence* is a <u>list</u> of numbers, and an *infinite sequence* is an infinite list of numbers.

General form of a Sequence

Infinite number of terms

$$a_1, a_2, a_3, a4, \ldots a_n^*$$

*   Term a_n is referred as "a sub n," and $\mathbf{a_n} = \dfrac{1}{\mathbf{2^n}}$, and

the *sequence* $\left\{ \dfrac{\mathbf{1}}{\mathbf{2^n}} \right\} = \dfrac{1}{2}, \dfrac{1}{4}, \dfrac{1}{8}, \dfrac{1}{16}, \ldots, \dfrac{1}{2^n}$

A series is an <u>addition</u> of the infinite number of terms of a *sequence*:

$$\frac{1}{2} + \frac{1}{4} + \frac{1}{8} + \frac{1}{16} + \ldots$$

Infinite Sequence

$$\{a_n\}$$

a_n means "in the limit", and the terms of the *sequence* in the limit get smaller and smaller; and if it goes out far enough, the term can get *close* to zero. As n approaches *infinity* **but never gets there**, an gets *closer* and *closer* to zero.

$$\lim_{n\to\infty} a_n = \lim_{n\to\infty} \frac{1}{2^n} + \frac{1}{2^\infty} + \frac{1}{\infty} = 0$$

Points on the *curve* $f(x)=\dfrac{1}{2^x}$ **make up** the *sequence* $\left\{\dfrac{1}{2^n}\right\}$

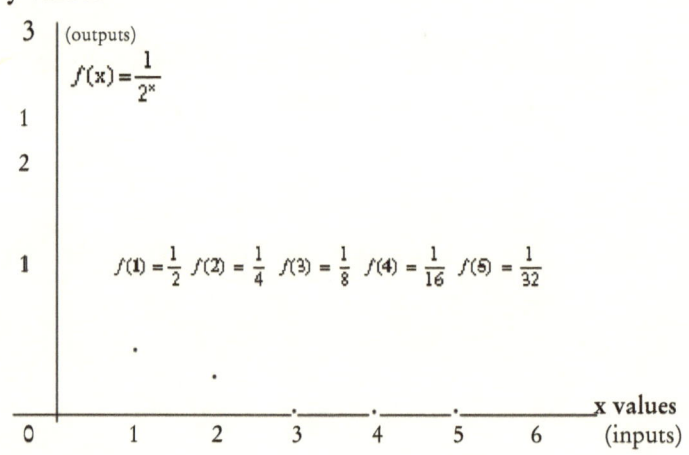

y values

3 | (outputs)

$f(x) = \dfrac{1}{2^x}$

1

2

1      $f(1)=\frac{1}{2}$ $f(2)=\frac{1}{4}$ $f(3)=\frac{1}{8}$ $f(4)=\frac{1}{16}$ $f(5)=\frac{1}{32}$

0   1   2   3   4   5   6   x values (inputs)

Determining Limits of Sequence

Does the *sequence* $a_n = \dfrac{n^2}{2^n}$ **converge or diverge?**

1. **Plug 1, then 2, then 3**, and so on into $\dfrac{n^2}{2^n}$ to generate first few terms:

$$\frac{1}{2},\ 1,\ \frac{9}{8},\ \frac{25}{32},\ \frac{36}{64},\ \frac{49}{128},\ \frac{64}{256},\ \ldots \qquad *$$

* $\dfrac{n^2}{2^n} = \dfrac{1^2}{2^1} = \dfrac{1}{2},\ \dfrac{n^2}{2^n} = \dfrac{2^2}{2^2} = \dfrac{4}{4} = 1,\ \dfrac{n^2}{2^n} = \dfrac{3^2}{2^3} = 9$, and so on.

2. **Use the L'Hopital's Rule and substitute**

(limit of function rule: $: f(x)\ \dfrac{x^2}{2^x}$)

L'Hopital's Rule

$$\lim_{x \to 0} \frac{f(x)}{g(x)} = \lim_{x \to 0} \frac{f'(x)}{g'(x)}$$

$$\lim_{x \to \infty} \frac{x^2}{2^x} = \lim_{x \to \infty} \frac{2x}{2^x \ln 2}\ \lim_{x \to \infty} \frac{2}{2^x \ln 2 \ln 2} = \frac{2}{\infty} = 0$$

The sequence $\dfrac{n^2}{2^n}$ *converges to zero.*

Summing up Infinite Series

An *infinite series* is simply adding up the *infinite* number of terms of a *sequence.*

$$\frac{1}{2^n} = \frac{1}{2},\ \frac{1}{4},\ \frac{1}{8},\ \frac{1}{16},\ \ldots,\ \frac{1}{2^n}$$

This *infinite* sum: $\frac{1}{2}+\frac{1}{4}+\frac{1}{8}+\frac{1}{16}+\ldots$ can be compacted as follows:

$$\sum_{n=1}^{\infty}\frac{1}{2^n}$$

The *summation symbol* $\sum$ indicates to plug **1** in for **n**, then **2**, then **3**, and so on. Also, an *infinite* sum is technically a *limit:*

$$\sum_{n=1}^{\infty}\frac{1}{2^n} \quad = \quad \lim_{b\to\infty}\sum_{n=1}^{b}\frac{1}{2^n}$$

nth Term Test

Convergence: Divergence:

$\lim_{n\to\infty} a_n = \mathbf{0}$ then $\sum a_n$ *converges* $\lim_{n\to\infty} a_n \neq \mathbf{0}$ then $\sum a_n$ *diverges.*

If the *terms* of the series   <u>converge to **zero**,</u> then the series <u>converge.</u>
If the terms of the series <u>do not converge to **zero**</u>, then the series <u>diverge.</u>

Three Basic Series

<u>Geometric Series:</u>

$$\sum_{n=0}^{\infty} ar^n = a + ar + ar^2 + ar^3 + ar^4 + \ldots{}^*$$

*   The first term, a, is called the "leading term," and r is called the ratio.
      For example,

Example: If a = 5 and r = 3

$$= 5 + 15 + 45 + 135 + \ldots$$

Note: ar^2 is $r^2 = 9$ and $5 \times 9 = 45$.

If a = ½ and r = ½,

$$= \frac{1}{2} + \frac{1}{4} + \frac{1}{8} + \frac{1}{16} + \frac{1}{32} + \ldots$$

Geometric Series Rule: If $0 < |r| < 1$, the geometric series *converges* to $\dfrac{a}{1-r}$.

If $\quad |r| \geq 1$, the geometric series *diverges*.

p-Series:

$$\sum_{n=1}^{\infty} \frac{1}{n^P} = \frac{1}{1^P} + \frac{1}{2^P} + \frac{1}{3^P} + \frac{1}{4^P} + \ldots$$

Note: It is called a *harmonic series* when p = 1 in the p-series, and it *diverges to infinity*. For example,

$$\frac{1}{1} + \frac{1}{2} + \frac{1}{3} + \frac{1}{4} + \ldots$$

p-Series Rule: If p > 1, the p-series *converges*.
                          If p ≤ 1, the p-series *diverges*.

Telescoping Series

$$\sum_{n=1}^{\infty} \left(\frac{1}{n} - \frac{1}{(n+1)} \right) = \left(1 - \frac{1}{2} \right) + \left(\frac{1}{2} - \frac{1}{3} \right) + \left(\frac{1}{3} - \frac{1}{4} \right) + \left(\frac{1}{4} + \frac{1}{5} \right) + \ldots$$

Note: The sum is $1 - \dfrac{1}{n+1}$, and in the *limit* as n approaches infinity $\dfrac{1}{n+1}$ *converges* to zero. Therefore, the sum *converges* to 1 + 0 or **1.**

Telescoping Series Rule:

    If $\dfrac{1}{n+1}$ *converges* to *finite number,* series *converges.*

    If $\dfrac{1}{n+1}$ *diverges,* series *diverges.*

Bibliography

Boles, Martha, and Rochelle Newman. *The Golden Relationship*. 1990

Conway, John H., and Richard K. Guy. *The Book of Numbers*. 1996

Devlin, Keith. *The Language of Mathematics: Making the Invisible Visible*. 1998

Durando, Furio. *Ancient Greece: The Dawn of the Western World*. 1997

Ghyka, Matila. *The Geometry of Art and Life*. 1977

Guant, Bonnie. *Beginnings: The Sacred Design; A Search for Beginnings, and the Eloquent Design of Creation*. 2000

Guedj, Dennis. *Numbers the Universal Language*. 1996

Hawking, Stephen. *The Illustrated on the Shoulder of Giants: The Great Work of Physics and Astronomy*. 2004

Herz-Fischler, Roger. *A Mathematical History of the Golden Number*. 1987

Huntley, H. E. *The Divine Proportion: A Study in Mathematics Beauty*. 1970

Ifrah, Georges. *The Universal History of Numbers: From Prehistory to the Invention of the Computer*. 2000

Kaku, Michio. *Hyperspace: A Scientific Odyssey Through Parallel Universes*. 1994

Lemesurier, Peter. *Decoding the Great Pyramid*. 1999

Livio, Mario. *The Golden Ratio: The Story of Phi, the World's Most Astonishing Number*. 2002

Mankiewicz, Richard. *The Story of Mathematics*. 2000

McDermott, Bridget. *Decoding Egyptian Hieroglyphs: How to Read The Secret Language of the Pharaohs*. 2001

Michel, John. *The Dimensions of Paradise*. 2001

Oakes, Lorna, and Lucia Gahlin. *Ancient Egypt: An Illustrated Reference to Myths, Religions, Pyramids, Temples of the Lands of the Pharaohs*. 2002

Pappas, Theoni. *The Joy of Mathematics*. 2004

Purce, Jill. *The Mystic Spiral*. 1974

Rawson, Philip. *Tantra: The Indian Cult of Ecstasy*. 1973

Schmmel, Annemarie. *The Mystery of Numbers*. 1993

Schneider, Michael. *A Beginner's Guide to Constructing the Universe: The Mathematical Archetypes of Nature, Art, Science*. 1995

Sigler, L. E. *Fibonacci's Liber Abaci: Leonardo Pisano's Book of Calculation*. 2003

Struik, Dirk J. A. *A Concise History of Mathematics*. 1987

Sutton, Daud. *Platonic and Archimedian Solids: The Geometry of Space*. 2002

Annex

TEN IMPORTANT GRAPHS

Graph 1 depicts an area divided into different size fractions.

Graph 2 depicts the x-y coordinate system that serves as a window into the world of calculus.

Graph 3 depicts a unit circle beginning at the positive x-axis showing in detail a 45° angle.

Graph 4 depicts in detail a tangent line and a secant line.

Graph 5 shows how the Mean Value Theorem for Derivatives works in graphic form.

Graph 6 depicts inverse functions, each is the mirror image of the other, reflected over the line y = x.

Graph 7 depicts a logarithmic function, which is *simply* an exponential function with the x and y axes switched. Namely, the up-and-down on an exponential function corresponds to the right-and-left direction on a logarithmic function, and the right-and-left direction on an exponential function corresponds to the up-and-down direction on a logarithmic function. Both of these functions are graphed on the same set of axes, so the relationship can be appreciated.

Graph 8 depicts different logarithmic functions.

Graph 9 depicts the family of curves $x^2 + C$.

Graph 10 depicts integrating $f(x)$, which means finding the area under the curve.

Graph 1

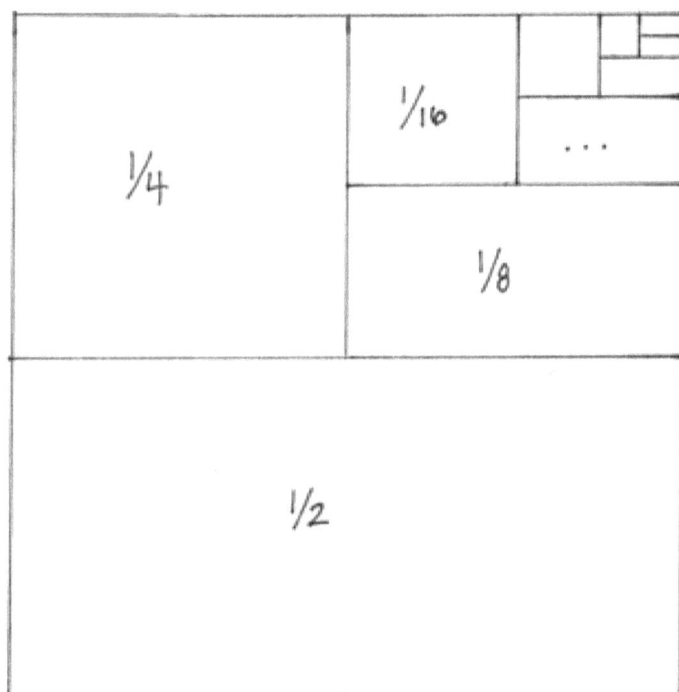

Graph 2

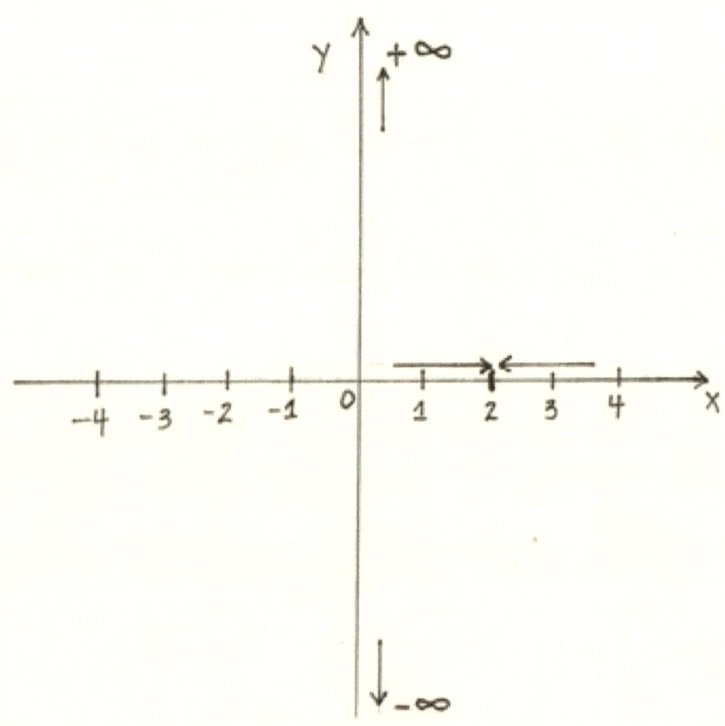

Graph 3

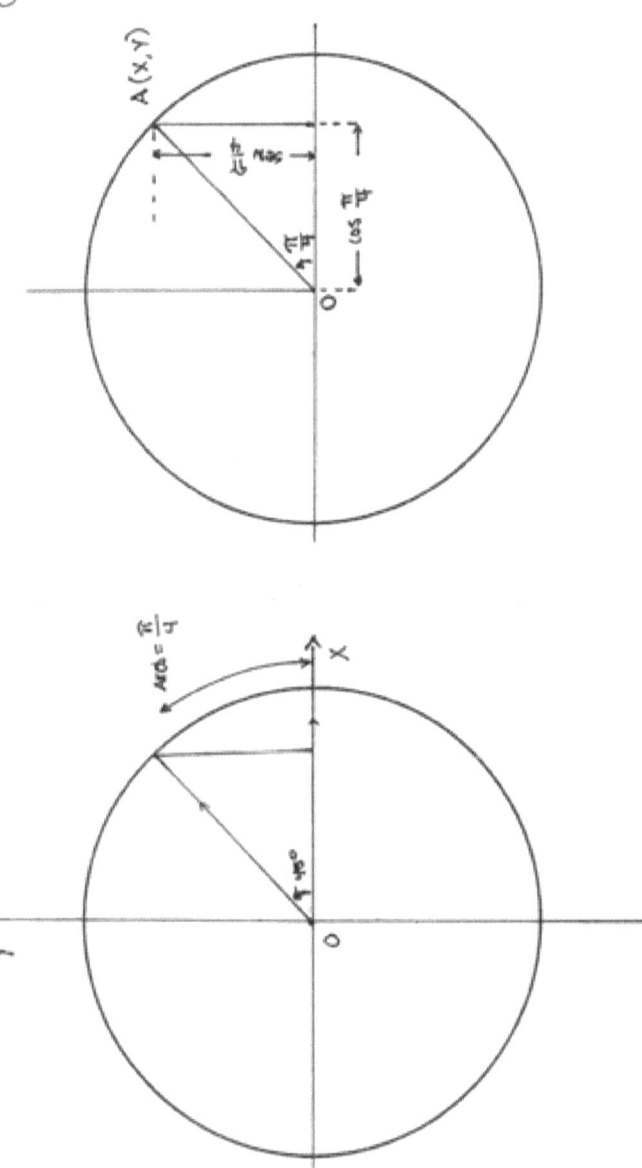

Graph 4

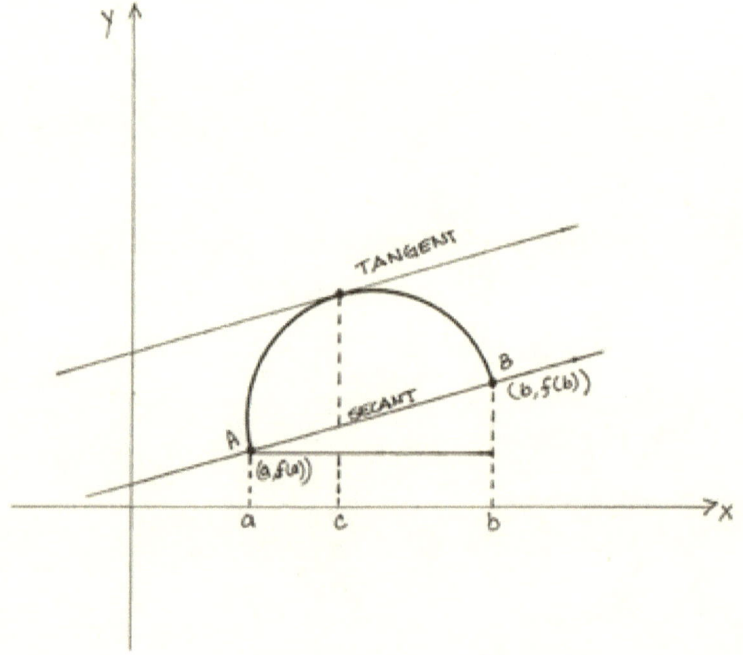

Graph 5

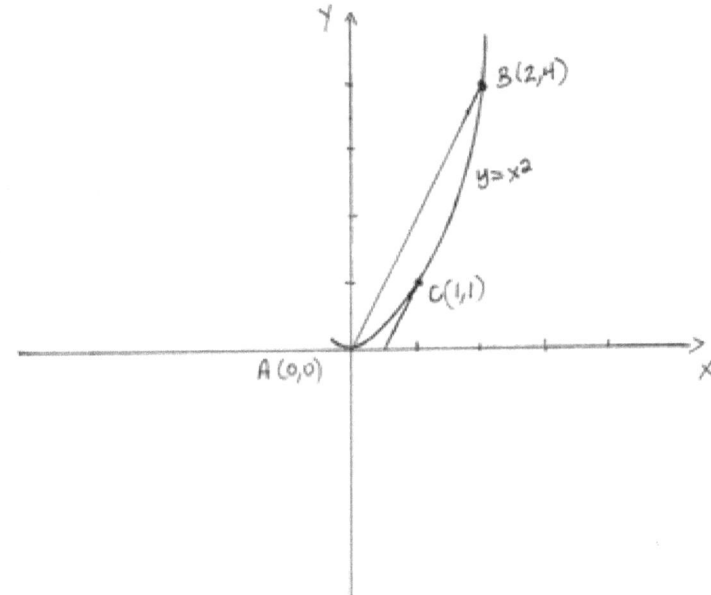

Y

B(2,4)

y=x²

C(1,1)

A (0,0)

X

Graph 6

(1)

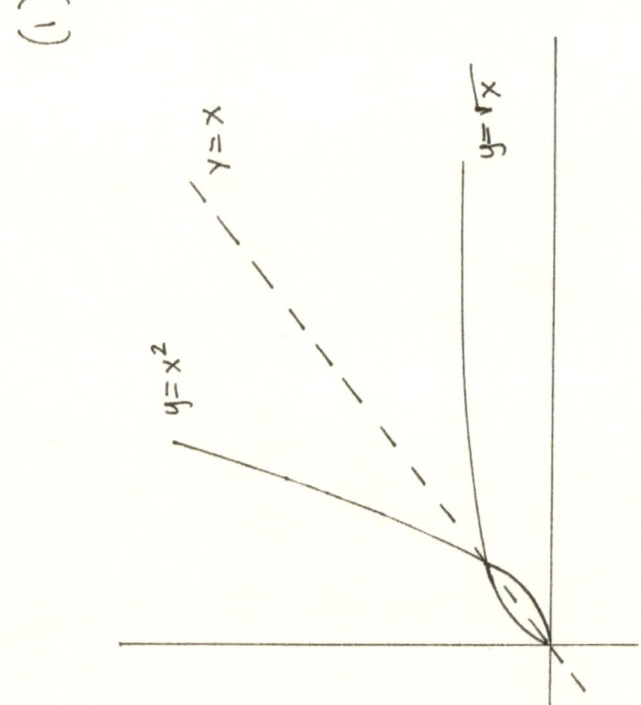

$y = x^2$

$y = x$

$y = \sqrt{x}$

Graph 7

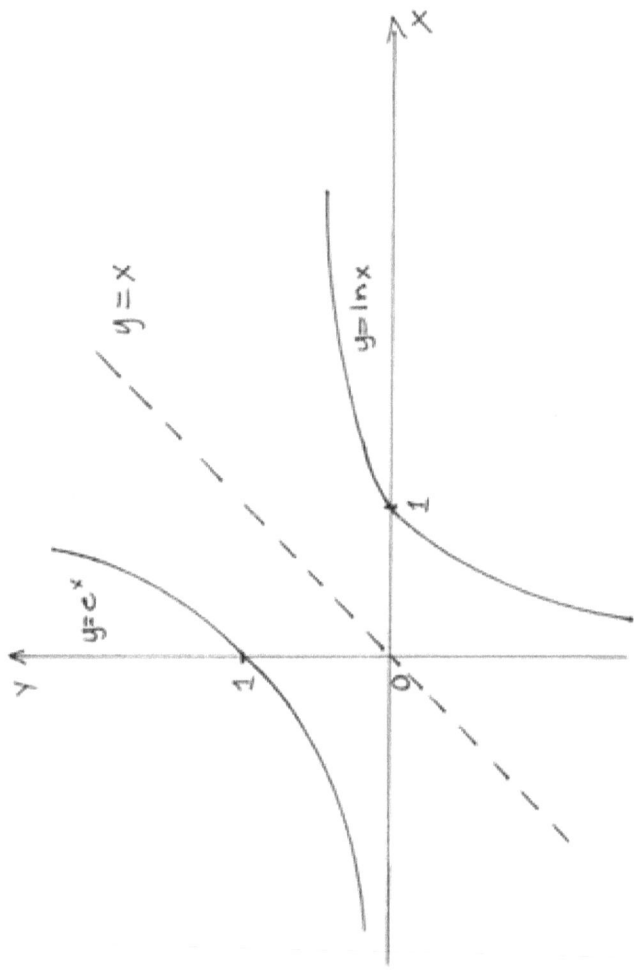

Graph 8

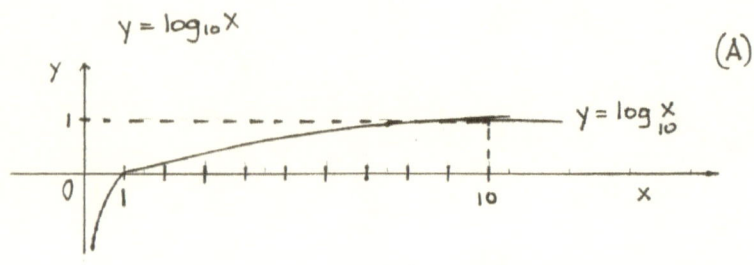

$y = \log_{10} x$ (A)

$y = \log \frac{x}{10}$

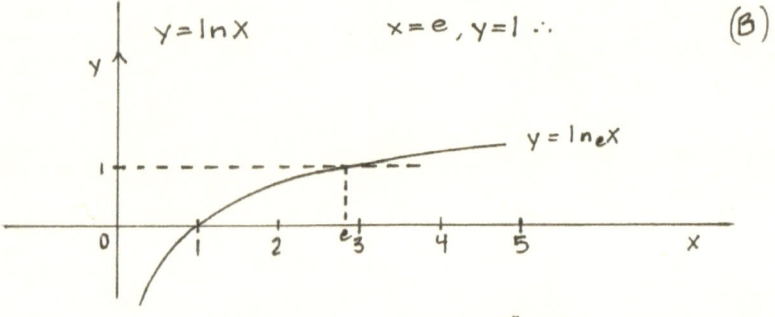

$y = \ln x$ $x = e, y = 1 \therefore$ (B)

$y = \ln_e x$

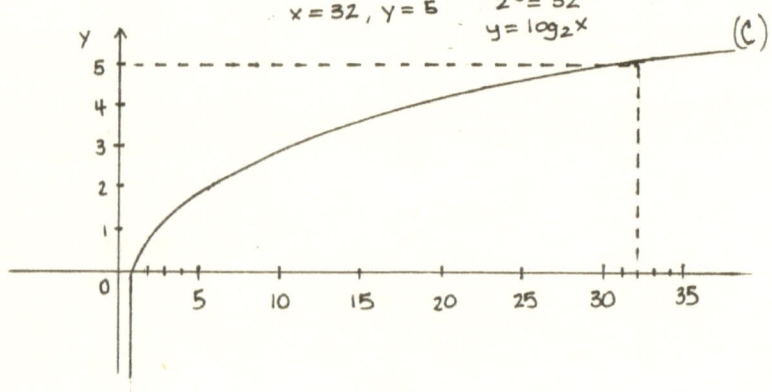

$x = 32, y = 5$ $2^5 = 32$

$y = \log_2 x$ (C)

Graph 9

$y = x^2 + C$

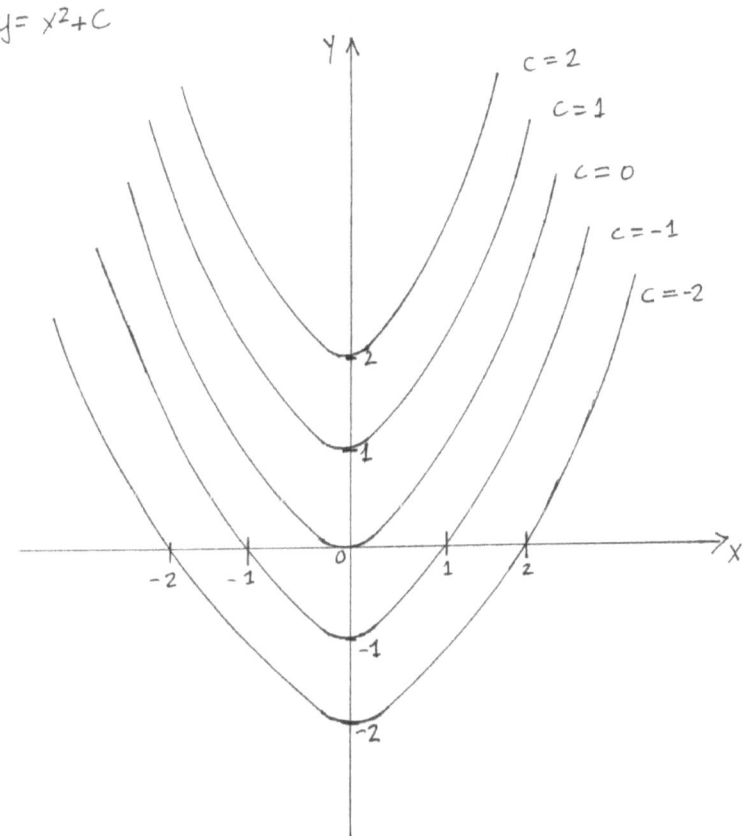

Graph 10

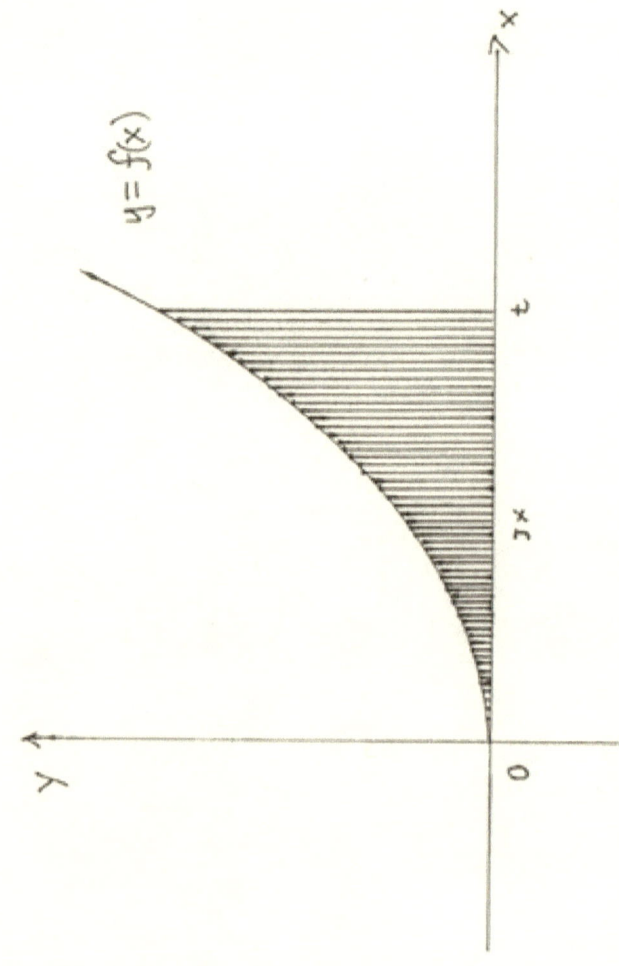

Index

About the Author

*O*livia F. Snyder has been an indefatigable fighter for the self-liberation of women and humanity, as well as the preservation of the earth's ecological system. At the same time she has been engaged in holistic self-healing for several years and has recently written several books on all of these subjects.

Snyder participated as speaker in the First World-wide Conference of the International Year of the Woman held in Mexico City. She has also worked for many years in diplomatic circles and international organizations, being actively involved in humanistic and women's issues in different countries around the world.